I like there to be someone in the historia
who tells the spectators what is going on,
and either beckons them with his high hand to look,
or with ferocious expression and forbidding glance
challenges them not to come near,
as if he wished their business to be secret.

—Leone Battista Alberti

■

ROGER SHATTUCK

FORBIDDEN KNOWLEDGE

FROM PROMETHEUS TO PORNOGRAPHY

■

A HARVEST BOOK
HARCOURT BRACE & COMPANY
San Diego New York London

This Harvest edition published by arrangement with St. Martin's Press.

Library of Congress Cataloging-in-Publication Data
Shattuck, Roger.
Forbidden knowledge: from Prometheus to pornography/Roger Shattuck.—
1st Harvest ed.
p. cm.—(A harvest book)
Originally published: New York: St. Martin's Press, 1996.
Includes bibliographical references and index.
ISBN 0-15-600551-4
1. Knowledge, Theory of, in literature. 2. Sex in literature.
3. Literature and morals. I. Title.
PN56.K64S43 1997
809'.9338—DC21 97-21658

Book design by Pei Loi Koay
Text set in Caslon 540
Printed in the United States of America
First Harvest edition 1997
C E F D B

For Nora

and to the memory of

Tari Elizabeth Shattuck

1951–1993

All men possess by nature a craving
for knowledge.

—Aristotle, *Metaphysics*

Forbede us thyng,
and that desiren we.

—Chaucer, *The Wife of Bath's Tale*

Individuum est ineffabile.

—Goethe, in a letter to Lavater

Parents and teachers should be aware that Chapter VII does not make appropriate reading for children and minors.

CONTENTS

■

FOREWORD

■

1.

Are there things we should *not* know? Can anyone or any institution, in this culture of unfettered enterprise and growth, seriously propose limits on knowledge? Have we lost the capacity to perceive and honor the moral dimensions of such questions?

Our increasingly bold discoveries of the secrets of nature may have reached the point where that knowledge is bringing us more problems than solutions. Contrasting threats like overpopulation and AIDS appear to be traceable to the effects of "progress." One powerful reading of history points out that the most advanced nations on Earth have produced unthinkable weapons of destruction at the same time as they have developed a media culture that revels in images of destructive violence. Can such a combination fail to propel us toward barbarism and self-annihilation?

In contrast, our most truly miraculous accomplishments as human beings take place unwittingly and privately, far removed from laboratories and studios and electronic screens, almost in another universe. For we learn to do certain things before we know what we are doing and in ways that no one can adequately explain. In twenty-four months, an infant learns to recognize and discriminate the elements of the world around it, learns to pull itself erect and to walk, learns to hear language and to talk. Is it possible that we accomplish these feats better for our lack of knowledge about how we do them? Can we know anything unwittingly? To ask the questions does not demonstrate that one has become a know-nothing and a Luddite. Proverbs in every language tell us that it is possible to know too much for our own good. Many great myths and legends explore the perils of knowledge. Fortunately, infants continue to learn to walk and talk. But many of us feel apprehensive about the future of our booming culture.

These exploratory remarks provide one path into my subject. I do not believe they exaggerate the picture. We have finally waked up to the dangers to our physical environment brought about by the depredations of human beings. But we have taken less notice of potential threats to our intellectual, artistic, and moral environments. It is to those three areas that I shall be referring constantly.

Another path into my subject leads more tranquilly through stories of people, ordinary and extraordinary, as they have responded to the world around them. This path leads to challenging tales of men and women whose lives still affect our own. Before going further back into the past, let me begin in the mid-nineteenth century.

One Victorian matron, a bishop's wife, became famous for the remark she made about evolution. She made it not so much about Darwin's circumspect *The Origin of Species* (1859) as about T. H. Huxley's belligerent *Man's Place in Nature* (1863). Darwin's young champion sang right out that man is "separated by no greater structural barrier from the brutes than they are from one another." When she heard the news, the matron displayed perfect cultural pitch: "Descended from the apes! My dear, let us hope that it is not true, but that if it is, let us pray that it will not become generally known" (Milner, 261: see the bibliography for complete references).

The matron wished to oppose the march of science and, if necessary, to quash an unsavory truth. We mock her squeamishness because we feel confident that nothing should stand in the way of the pursuit and communication of knowledge. But the lessons of history and the nature of contemporary events do not always support that confidence. The matron's naïve response reveals an anxiety we cannot simply dismiss as baseless prejudice. She articulated a rudimentary understanding of forbidden knowledge.

In every age, news of wars and disasters and crimes has been appalling. Without overcoming those ancient woes, we now have new ones to lament. In the late twentieth century, we reckon with reports of marvels, which are also afflictions, brought about not by backwardness and ignorance but by advancing knowledge and its applications. Not only the most barbarous nations but also the most civilized expend vast resources to develop nuclear and biological weapons of unthinkable destructive force. Genetic research raises the remote prospect of choosing our children's physical and mental endowment like wallpaper patterns. The invasive presence of audiovisual media in our lives from earliest infancy threatens to shape our character and behavior as forcefully as genetic manipulation. In our quest for energy sources, we may be reducing the life span of our planet. Scientific research, freedom of speech, the autonomy of art, and academic freedom combine forces, as I shall argue in Part Two, to carry us beyond our capacities as human agents to control our fate. Our greatest blessings confound us.

This great wrestling match with the best-endowed and most advanced parts of ourselves was dramatized in deeply opposing ways by two modern works whose importance has increased with time, *Alice's Adventures in Wonderland* (1865) and *The Strange Case of Dr. Jekyll and Mr. Hyde* (1886). Both belong to the era of British technological and commercial leadership in the world and of a Western imperialism that mixed exploitation with philanthropy. In the middle of the Victorian era, Lewis Carroll peered into the dreamworld of an adolescent girl and found it peopled with grotesque creatures making strange demands of her good intentions. Nothing goes quite right, and nothing goes irretrievably wrong. Alice suffers no harm and wakes up having learned that the creatures within us are essentially benign under their fearsome eccentricities. And beneath Carroll/Dodgson, the friend and photographer of prepubescent girls, one finds not a child molester but a chaste poet of nature's riddles and paradoxes. Alice masters her fears and returns home to a secure existence.

Robert Louis Stevenson offers us a totally different vision of the world. Friends of Dr. Jekyll, a respected London physician with a penchant for unorthodox medical research, discover that he is lodging a suspicious scoundrel in an apartment connected to the rear of his own house. This mysterious Mr. Hyde commits a horrible murder and disappears. Several months later, Hyde is found in his apartment, dead by his own hand. Jekyll has disappeared. Jekyll's full confession in writing finally solves the mystery in the last chapter.

Jekyll and Hyde are two elements, two contrasting outward forms of one person. The doctor has discovered a drug that proves the duality of human life. The drug transforms Dr. Jekyll into his purely evil self, Mr. Hyde, in whose person he can pursue "undignified pleasures" into undescribed excesses. Another potion is needed to suppress Hyde and to restore the doctor to his usual human existence. Finally, the monstrous persona of Hyde gains the upper hand. Drugs can no longer suppress Hyde and reaffirm Jekyll for more than a short time. And even those medicines are running out. In his last written words before surrendering to his evil self, Jekyll attempts to disclaim responsibility for the monster he has produced out of himself and to deny that the evil of Hyde has besmirched the soul of Jekyll.

Beginning as a fairly sedate mystery story, Stevenson's fifty-page novella soon turns into a full-fledged horror tale with suggestions of vampirism and superhuman powers intervening with the help of secret substances to transform the balance of life itself. Stevenson's moral fable, based both on a vivid nightmare and on newspaper accounts of an Edinburgh businessman-thief, seizes our imagination from two sides. First, we respond with some sympathy to the figure of the double, the respectable citizen fettered to a depraved alter ego. In this era of growing anonymity and nomadism and of hypnotic media images feeding an alluring fantasy life within, we are constantly encouraged to develop a covert life of violent excesses. From this point of view, the fable records not a "strange case" but the common temptation to lead two lives. Second, we respond with apprehension to the figure of the fanatic doctor who has cracked the secrets of life and human identity. His truly "strange case" frightens us because of the destruction his experiments let loose upon ordinary citizens. Furthermore, the story implies that Dr. Jekyll's struggle is not so much with the conventional embodiment of evil called Hyde as with his own higher knowledge and mysterious powers. Dr. Jekyll discovers evil by succumbing to the allurements of his own genius.

Most of us have welcomed both Alice and Dr. Jekyll into our fantasy life. Alice reassures us. Dr. Jekyll, in contrast, carries us into an ominous dilemma, the confrontation of truth and its consequences. For, through experiments on his own person, the obsessed scientist demonstrates that the truth may have unforeseen and devastating consequences. The evident dangers of his experiments lure him on rather than restraining him. Jekyll's gifts in the pursuit of truth unstring his moral character.

The Victorian matron who did not want to know the truth and Dr. Jekyll, who could not hold himself back from toying with the most dangerous and compromising forms of truth, provide my first two parables of forbidden knowledge.

Taboo, occult, sacred, unspeakable—with such terms, earlier cultures recognized limits on human knowledge and inquiry. What has happened to the venerable notion of forbidden knowledge? In the practicalities of daily living, we accept constraints ranging from environmental regulations to truancy laws to traffic lights. In matters of the mind and its representations, Western thinkers and institutions increasingly reject limits of any kind as unfounded and stultifying. We have outgrown the need to punish heresy and blasphemy. Both scientific research and the worlds of art and entertainment rely on an unspoken assumption that total freedom in exchanging symbolic products of mind need not adversely affect the domain of daily living and may well enhance it.

On the one hand, we have laws and customs to limit behavior, though often trampled by scoff-laws, violent gangs, and organized crime. On the other hand, the symbolic products of mind—words, images, movies, recordings, television shows—do not and presumably should not fall under similar restraints. That divergence furnishes the essential dynamic of Western culture in its long history of expansion in all realms. And that divergence merits close scrutiny.

2.

> *Socrates: "All things are knowledge, including justice, and temperance, and courage—which tends to show that virtue can certainly be taught."*
>
> —PROTAGORAS

> *Ye shall know the truth and the truth shall make you free.*
>
> —JOHN 8:32

Will knowledge solve our problems? Will an "explosion" of knowledge reduce hardship among us and make us just, virtuous, and free? History suggests that the West has accepted this optimistic wager, though not without doubts and forebodings. We believe that the free cultivation and circulation of ideas, opinions, and goods through all society (education, scholarship, scientific research, commerce, the arts, and the media) will in the long run promote our welfare. We also believe that we can contain the social and political upheavals into which these same cultural enterprises have launched us. At the end of the second millennium C.E., I believe we have arrived at a crisis in our lengthy undertaking to reconcile liberation and limits.

The two quotations above invoke *knowledge* and *truth*. But Socrates' and Jesus' words do not sit well today in a society that tends to doubt rather than to honor traditional knowledge and in which many educated skeptics snicker at the word *truth*. During the three hundred years since the Enlightenment, we have made life difficult for ourselves precisely in the domains of knowledge and truth. Having to a large extent dismissed any faith in revealed or absolute knowledge, how can we distinguish the true from the untrue? And while we seek empirical or pragmatic means to do so, another question, both larger and more precise, looms before us. Can we decide if there are any forms of knowledge, true or untrue, that for some reason *we should not know*?

In the poem "The Oxen," Thomas Hardy retells an ancient English folktale about farm animals kneeling in their stalls at midnight of the Nativity.

Christmas Eve, and twelve of the clock,
"Now they are all on their knees,"
An elder said as we sat in a flock
By the embers in hearthside ease.

It is puzzling that Hardy's poem omits the most arresting part of the tale. *Any person who goes to the stable to verify the truth of the fanciful story will die before the year is out*—presumably regardless of what one finds in the stable. Don't peek. Leave well enough alone. Here is a quandary for believers. Does doubt corrupt or enlighten? Does faith survive best on ignorance or on knowledge? Need we verify all traditional beliefs by rational inquiry? Hardy's farmers do not think so.

A familiar tale for children treats curiosity with greater sympathy. Rudyard Kipling, bard of the British Empire and world traveler, distilled his keen understanding of human nature and his wry sense of humor in *Just So Stories for Little Children* (1902). "The Elephant's Child," from that collection, describes a well-brought-up young elephant who finally gets tired of the spankings he receives from all his family. On the advice of the Kolokolo Bird, he decides to go away to "the great grey-green, greasy Limpopo River, all set about with fever-trees" to satisfy his inquisitiveness about what the Crocodile has for dinner. At this time, elephants have a snout only, no trunk. When he steps unexpectedly on a Crocodile, the Elephant's Child is caught by the snout and almost pulled under water by his powerful adversary. The Bi-Coloured-Python-Rock-Snake, to whom the Elephant's Child has been very polite, saves his new friend. In the great tug-of-war, the stubby snout is stretched out into a multipurpose trunk. The Elephant's Child uses it to establish his authority when he returns home, and many relatives go off to obtain their own nose job. Elephant culture has been greatly enhanced by the youngster's expedition. Kipling's good-natured story leaves the impression that, if one has been properly brought up to respect others, curiosity (or "'satiable curtiosity" in Kipling's well-tuned malaprop) has advantages that outweigh its risks.

We would do well to be attentive to such tales. Today we recognize virtually no constraints on our freedom and our right to know. Is curiosity the one human drive that should never be restricted? Or does it embody the greatest threat to our survival as ourselves? Kipling answered with a jolly parable that counsels curiosity within limits. The term *forbidden knowledge* takes a harsher approach to these questions. It represents a category of thought with a long history, too complex to be one of Lovejoy's "unit ideas," yet demonstrably the armature of many powerful narratives.

In the pages that follow, I propose an inquiry into forbidden knowledge—an inquiry with an outcome, not a theory of forbidden knowledge with illustrations.

3.

The two atomic bombs dropped on Japan in August 1945 probably saved my life. At least I long believed that statement. After a year in the Southwest Pacific as a combat pilot, I had been assigned to a bomb wing in Okinawa that was staging to go ashore in the first wave of landing craft invading mainland Japan. We had the mission of opening an airstrip near the beachhead. We were not told where the invasion would take place. We were told very clearly to expect more than 50 percent casualties. Then early one evening, the PA system hanging over the pyramidal tents came to life with a mysterious message about "a new kind of bomb" and a city named Hiroshima. Someone in the camp yelled, "The war's over."

A few weeks later, when we had become liberators of Korea instead of invaders of Japan, I flew a B-25 up the Inland Sea of Japan to have a look at Hiroshima. From a thousand feet in the shattering silence of the cockpit, we could see a flattened smoldering city. We did not know the number and nature of casualties and the intensity of radiation we were foolishly flying through. We learned about all that a year later from the issue of *The New Yorker* devoted to John Hershey's *Hiroshima.*

Fifteen years later, no one could mistake the global consequences of the two bombs. The world was locked in "the balance of terror." On Easter Sunday 1961, I joined a three-hour march from the capitol steps in Austin, Texas, to Bergstrom Strategic Air Command Base to demonstrate against the manufacture and deployment of nuclear weapons. From passing cars and trucks, people spat at us and threw beer bottles. But my convictions were unalloyed by doubts.

Marked by that series of events, I have lived out my biblical portion of years with a warning light constantly flashing in my peripheral vision. It continues to signal that we have strayed off course, that some mechanism has malfunctioned. How could so human a President as Truman have made the decision to drop two atomic bombs on heavily populated areas? How could we go on to endanger more lives and whole societies by developing the hydrogen bomb? And then by what perverse human logic did those unthinkable weapons succeed in keeping the peace between two enemy superpowers for almost half a century? As the millennium approaches, are we—not just we Americans but we citizens of the Earth—losing control of our future because of the threat of nuclear proliferation?

The warning light still flashes. I have come to believe that its signal refers not only to the destructive forces we have conjured out of the atom but even more essentially to a condition we have lived with always: the perils and temptations of forbidden knowledge.

This book has a personal origin. But I shall leave autobiography behind in order to pursue my subject in what I believe to be the best set of records we have about ourselves: *stories* of all kinds, true, embellished, invented. We are often taught to deal with *ideas* as the highest form of knowledge. But the process of abstraction by which we form ideas out of observed experience eliminates two essential aspects of life that I am unwilling to relinquish: time and individual people acting as agents. At their purest, ideas are disembodied and timeless. We need ideas to reason logically and to explore the fog of uncertainty that surrounds the immediate encounter with daily living. Equally, we need stories to embody the medium of time in which human character takes shape and reveals itself to us, and in which we discover our own mortality. More than a history of ideas, these pages offer a history of stories.

4.

Part One of this book deals with literary works. The first chapter assembles a large diversity of materials in order to sketch out the dimensions of the subject. These are the most demanding pages. Each of the following four chapters concentrates on only one or two works. Part One provides an overall history of forbidden knowledge and a substantial sampling of its varieties.

Part Two deals with two contemporary questions: the challenge of science and the problem of pornography as represented by the recent rehabilitation of the Marquis de Sade. The final chapter considers the practical and moral implications of forbidden knowledge and their significance for our future. Some readers may be tempted to turn directly to Part Two. In that case, I hope they will subsequently return to Part One, for the works there discussed provide a pertinent background for the urgent problems raised in Part Two.

Part One

LITERARY NARRATIVES

■

Chapter I

THE FAR SIDE OF CURIOSITY

■

A few years ago a meeting of prominent scientists and science writers in Boston devoted a session to discussing what motives had brought them to the pursuit of science. All in the group (it included Isaac Asimov, Freeman Dyson, Murray Gell-Mann, and Gunther Stent) cited curiosity about the workings of the world as the fundamental factor. Fame, riches, truth, and the greater glory of God were not mentioned.

We have no historical records to inform us how or why human beings first began to find explanations for the great regularities in nature like animal migrations, the movements of sun, moon, and stars, and the seasons. But we surmise by an imaginative leap and from a few prehistoric cave drawings that instincts of self-defense and survival were equaled by an impulse of idle curiosity—like that of the Elephant's Child. At least a few cavemen wanted to know more than was necessary for their immediately foreseeable needs. As organized societies developed, curiosity became particularly strong at crucial periods like sixth-century Greece, the Italian

Renaissance, and the northern Enlightenment. Like poverty, curiosity we have always with us.

In order to discover the sources of forbidden knowledge and how it occupies a place close to curiosity at the center of Western culture, I shall start with Greek mythology and Old Testament stories. Both before and after these two fertile streams mingled into what we now name and number as the Common Era, they developed a pair of reciprocating attitudes toward knowledge: liberation and limits. I shall follow these attitudes through a selection of stories covering three millennia of human history.

1. PRESUMPTION: PROMETHEUS AND AFTER

Hesiod, who seems to have been a farmer-poet in eighth-century Boetia, gives us some of the best accounts of the Greek gods and their dealings with mortals. Embellishing traditional oral versions, he wrote two major sequences about Prometheus, a demigod who stole fire from Zeus in order to save men (still without women) from extinction. The wily Prometheus, a friend of mankind, tricked Zeus by withholding the best parts of a sacrificed ox. "That is why Zeus devised troubles and sorrows for men. He hid fire. But Prometheus, noble son of Iapetos, stole it back for man" (*Works and Days*, 49–51). Stung now in the depths of his being, Zeus bound Prometheus to a rock, with an attendant vulture to eat out his liver. The stolen gift of fire has been variously interpreted as representing a great number of crucial human capacities—mechanical arts, science, language, imagination, consciousness itself. Prometheus became our benefactor by making a raid on the knowledge withheld from us by Zeus in his anger. Prometheus' defiance became our salvation in an episode that appears to rebut the proverb that ignorance is bliss.

But it is unwise to deprive Prometheus of the rest of his story in Hesiod's versions. In retaliation for Prometheus' insubordination, Zeus sent Pandora, the first woman. She, too, was a gift, not stolen, but made to order to tempt Prometheus' gullible brother, Epimetheus. By falling victim to her charms, Epimetheus brought into our midst the female whose name means "giver of all" or "gift of all." What Pandora gave us, when she removed the lid of the jar or box the gods sent with her, is grief, cares, and all evil. Her curiosity about the contents of the jar matches Epimetheus' curiosity about a new companion, a modest maiden "with the mind of a bitch" (Hesiod). The dire effects of her "gifts" cancel out the benefits bestowed by Prometheus' defiance of the gods.

Now, later versions of the Prometheus story that have come down to us usually make no mention of the closely linked figure of Pandora. Prometheus' daring raid on Olympus produces a liberating fire for our ancestors, and the further consequences of that raid are forgotten. The most famous literary treatments of the Prometheus myth—a page in Plato's *Protagoras*, Aeschylus' *Prometheus Bound*, Shelley's *Prometheus Unbound*—leave out Pandora as an awkward appendage or complication. Thus they avoid dealing with the full consequences to humankind of the knowledge Prometheus brings as narrated in Hesiod's earliest versions. Here is another instance of truth, Prometheus' fire, being separated from its consequences, Pandora's disruptive presence among men. We may not like the full myth, but we are distorting it by cutting it in two. In classical Western painting, Pandora went on to become an allegorical figure for "beautiful evil."*

Even in its full version, the Prometheus and Pandora story does not fuse so dramatically as the Adam and Eve story does themes of knowledge, curiosity, sexuality, the origin of evil, and mortality. In Hebrew Scripture, however, no figure assumes the defiant role of Prometheus in Greek mythology, not Adam, not the shadowy personage called Satan, not even one of the prophets. A better case can be made (as Milton later did) for a parallel between Eve, by whom temptation and sin enter Eden, and Pandora, by whom all evils are brought down on

---

*At long intervals, Pandora receives attention on her own account. For twenty years at the beginning of the century, the German expressionist playwright, Frank Wedekind, rewrote his Lulu drama about a femme fatale and bitch goddess whose sexual appetite cuts a broad swath of corruption and murder through Victorian society. She ends up a common prostitute who is killed by Jack the Ripper. Wedekind's two Lulu plays, *Earth Spirit* (1895) and *Pandora's Box* (1903), allude to a fragmentary verse drama by Goethe, *Pandora's Return* (1818), and to the comparable figure of feminine evil painted in Zola's *Nana* (1880). Alban Berg chose Wedekind's Pandora dramas as the basis for his unfinished twelve-tone opera, *Lulu* (1937).

mankind. After these two profoundly human tales, sobering yet not without comic overtones, the theme of prideful curiosity never disappears from the history of Western culture.

I shall hold Adam and Eve for the following chapter. Even apart from them, Genesis and Exodus remain rich in stories related to forbidden knowledge. The familiar verses about the Tower of Babel recount another episode of pride and fall. It is almost impossible to overinterpret them. They raise themes of the city, of overweening ambition, of the dangers of technology, of the origin of languages, cultures, and races. Since the Flood, there had been only one people under Noah. After Babel, the Torah ceases tracing "the whole Adamic race," as the Scofield Bible phrases it, and devotes itself to "a slender rill"—the nation of Israel. This time, it is the Lord himself who opens the jar and releases over the earth confusion of tongues. I quote the entire passage.

> *And the whole earth was of one language, and of one speech.*
>
> *And it came to pass, as they journeyed from the east, that they found a plain in the land of Shinar; and they dwelt there.*
>
> *And they said one to another, Go to, let us make brick, and burn them thoroughly. And they had brick for stone, and slime had they for mortar.*
>
> *And they said, Go to, let us build us a city and a tower, whose top* may reach *unto heaven; and let us make us a name, lest we be scattered abroad upon the face of the whole earth.*
>
> *And the Lord came down to see the city and the tower, which the children of men builded.*
>
> *And the Lord said, Behold, the people* is *one, and they have all one language; and this they begin to do: and now nothing will be restrained from them, which they have imagined to do.*
>
> *Go to, let us go down, and there confound their language, that they may not understand one another's speech.*
>
> *So the Lord scattered them abroad from thence upon the face of all the earth: and they left off to build the city.*
>
> *Therefore is the name of it called Babel; because the Lord did there confound the language of all the earth: and from thence did the Lord scatter them abroad upon the face of all the earth.*
>
> (GENESIS 11:1–9)

In Eden, the Lord declares directly to Adam and Eve his prohibition against eating the fruit of the Tree of the Knowledge of Good and Evil. But no one has warned the citizens of Babel or Babylon that they must observe certain limits in their investigations of the world. They have discovered the new technology of bricks and mortar and put it to the inevitable use of building a tall tower. The Tree of Knowledge was set out by God, we speculate, as adornment and probation. The godless tower built by the Babylonians represents their wish for personal aggrandizement: "Let us make a name" (11:4). If this vainglorious project were to "reach unto heaven," God's majesty and mystery would be defiled. In punishment, the Lord does not destroy Babylon; he divides to conquer and of one people makes many with different customs and languages.

In these same verses about confounding the ambitions of humanity, a momentous faculty appears for the third time in the King James translation of Genesis: *imagination.* "And now nothing will be restrained from them which they have imagined to do" (11:6). United by technology and a universal language, humanity achieves untoward power. Power in itself does not endanger. But imagination linked to power may exceed the limits of the human condition and aspire to godhead.

We see it happen the first time before the Flood. "And God saw that the wickedness of man was great in the earth and that every imagination of the thoughts of his heart was only evil continually" (6:5). The Lord repents of his creation and finds Noah alone worthy of survival as "a just man." Two chapters later, after the Flood has receded, Noah's burnt offerings persuade the Lord not to destroy mankind again. But the verse contains the same demurrer and warning about the nature of man: "For the imagination of man's heart is evil from his youth" (8:21). Both passages point ahead to the Tower of Babel episode, in which the overheated imagination, the dark side of curiosity, calls down punishment on itself.* And as I read them, the three Old Testament passages establish the

* The original Hebrew does not disqualify the English. The term in Genesis at both 6:5 and 8:21 is *yatzer*, derived from a verbal root meaning "to shape" or "to fashion," as in the activity of a potter. "Devisings" is probably a more accurate version than "imagination." Ancient Hebrew was short on abstractions and terms for mental faculties. The verse at 11:6 on the Tower of Babel uses a different word, *yazam*, meaning "to plot, to conspire, to aspire." The word *yazam* carries a negative

link between curiosity and imagination that will recur in every chapter of this book.

Later events in the history of the nation of Israel as it moves from the Noahic covenant to the Abrahamic and the Mosaic covenants treat a further aspect of forbidden knowledge closely related to unbridled curiosity and imagination: Can anyone look upon the Lord? First Jacob: "And Jacob called the name of the place Peniel: for I have seen God face to face, and my life is preserved" (Genesis 32:30). Several generations later, during the trials of the escape from Egypt, Moses lives through a set of searing and contradictory encounters. Twice he succeeds as well as Jacob in seeing God (Exodus 24:10 and 33:11). The latter passage sweeps aside all ambiguity. "And the Lord spake unto Moses face to face, as a man speaketh unto a friend." But Scripture runneth in mysterious ways. Echoing several other passages (e.g., Moses hides his face from the Burning Bush [Exodus 3:6], a movement repeated in Exodus 19:12 and 19:21), the close of the same chapter reverses the situation dramatically. Even as the Lord declares that Moses has found grace in his sight, he sets out the rules and improvises a little scenario to illustrate them.

> *And he said, Thou canst not see my face: for there shall no man see me, and live.*
>
> *And the Lord said, Behold,* there is *a place by me, and thou shalt stand upon a rock:*
>
> *And it shall come to pass, while my glory passeth by, that I will put thee in a clift of the rock, and will cover thee with my hand while I pass by:*
>
> *And I will take away mine hand, and thou shalt see my back parts: but my face shall not be seen.*
>
> (EXODUS 33:20–23)

---

connotation in biblical Hebrew. (I am grateful to Robert Alter for providing this information.)

In the light of the original Hebrew, of the drift of meaning at these three points, and of our gradual understanding of ourselves as moral agents (which is my subject), I feel that the King James choice of "imagination" in all three places does not lead us astray. It represents a brilliant stroke in English translation, a justified leap of meaning consistent with the way Genesis shows certain inward inclinations of the human heart as leading us into trouble.

Writing on Exodus in *The Literary Guide to the Bible,* J. P. Fokkelman identifies "the main issue of the book: the question of whether man can behold God or not." By planting the Tree of the Knowledge of Good and Evil in the Garden of Eden, the Lord appears to issue to his new creatures a covert invitation to both companionship and rivalry with him. But we must remember that the apple did not divulge to Adam and Eve full knowledge of things, let alone of the Lord in his quiddity. In the *Purgatorio,* Dante insists on the point.

> *Content you with* quia* *sons of Eve;*
> *For had you power to see the whole truth plain*
> *No need had been for Mary to conceive.*
>
> (III, 37–39, TR. DOROTHY SAYERS)

Had the apple revealed everything to Adam and Eve, no further revelations would have been called for. The entire action of the two Testaments of the Bible and of subsequent history is predicated on the *partial* knowledge granted to the human mind and achieved by it. In these early books of Hebrew Scripture, the Lord seems to alternate among the roles of a beneficent Prometheus, a treacherous Pandora, and an awesomely stern Zeus.

Another haunting cluster of ancient stories from both Hebrew Scripture and pagan myth concerns a similar prohibition laid upon the human faculty of sight. In these tales, sight stands for the human need for evidence of the senses to bolster a flagging faith. The results are often fatal. Lot's wife, escaping the destruction of Sodom, hears the injunction, "Look not behind thee" (Genesis 19:17). When she turns to look at the horrible scene of fire and brimstone, "She became a pillar of salt" (Genesis 19:26). Her weakness of will closely parallels that of Orpheus leading Eurydice out of the underworld. In spite of instructions to the contrary, he must verify with his eyes that his wife has not faltered along the way. That failure of faith deprives him of Eurydice for the second time and for good.

But stories of ocular prohibition do not always end tragically.

---

*Finite knowledge of effects, not final knowledge of essences.

Told not to look at the horrible Gorgon's head of Medusa, Perseus obeys orders, escapes petrification by looking only at Medusa's reflection on his shield, and uses other magic accoutrements to behead the monster. He can contain whatever curiosity he feels to behold Medusa's ultimate ugliness directly, a temptation that might lead others of us to meet the fate of Lot's wife. Shem and Japheth lay a garment over their heads and look the other way when they go in to cover their father Noah's drunken nakedness. The apostle Thomas, who doubted Jesus' resurrection until he received ocular and tactile proof, went unpunished but was accorded a stern rebuke. "Blessèd are they that have not seen and yet have believed" (John 20:29). The number of doubting Thomases among us has grown very large.

An appealing variation on these events comes down to us in Apuleius' *The Golden Ass.* Another incarnation of Eve and Pandora, Psyche must bear the burden of beauty so great that it provokes Venus' jealousy. Venus' well-favored son, Cupid, instead of following his mother's instructions to make Psyche fall in love with a mean and ugly husband, falls in love with her himself. Through the intervention of the oracle, she is sequestered in a beautiful palace where Cupid can visit her at night without revealing his appearance and identity. Psyche is content for a time with her situation. Then, warned by her envious sisters that her lover may be a monster, she wishes to find out his true shape. The lamp by which she discovers Cupid's beauty while he sleeps lets fall on him a drop of hot oil that wakens him. He flees, murmuring, "Love cannot last without trust." Psyche now seeks Cupid everywhere, submitting to and surviving (with the help of nature's creatures) the cruel trials imposed on her by Cupid's still-jealous mother, Venus. The last trial sends Psyche to the underworld to fetch a box containing a token of Proserpina's beauty in order to restore Venus' splendor. Told not to pry into the box, Psyche again cannot repress her curiosity and her vanity. She peeps into the box and is immediately overcome by a Stygian sleep. The story ends happily when Cupid rescues Psyche, intervenes with Jupiter to have her immortalized as a goddess, and establishes their union in the heavens. Psyche twice destroys her potential happiness by wishing to know more than she should. Unlike Lot's wife and Orpheus, she is rescued by a loving god, who lifts her out of the human condition and presumably tries to cure her of curiosity. Milton, La Fontaine, Molière, Keats, César Franck, and innumerable painters have celebrated the story of Cupid and Psyche as a modernized and secularized version of Adam and Eve with a happy ending.*

Because it ends with Œdipus putting out his eyes in horror at what he has learned about himself, *Œdipus the King* presents itself as the extreme case of a character being punished for seeing what is forbidden. Yet Sophocles' tragedy will not quite fit. On the one hand, Œdipus is the innocent and ignorant victim of two fiendishly interlocking Delphic oracles that concern the two royal couples who, respectively, bore and raised him. How possibly can we blame Œdipus for anything? On the other hand, his full-blown Athenian character (overbearing, high IQ, prideful), goaded by the third oracle (about an assassin, to be found in Thebes, who is the cause of the plague), drives him to discover the facts that will devastate his and his family's life. Œdipus displays no freedom and no courage in seeking out the awful knowledge. By temperament and by divine intervention, he has no choice. He enacts his doom as contained in the oracles that hoodwink all parties, including him. The "tragedy" could have been avoided only if his character had been different (enough to prevent him from becoming so enflamed as to kill an old man in a wagon who claimed the right of way) or if the gods had stayed out of mortal affairs.

I am suggesting that whereas we think of Lot's wife, Perseus, Orpheus, and Psyche as having the freedom to choose their conduct, Œdipus is so entrapped in mysterious oracles and in the larger-than-life expectations of Athenian character that he simply follows his fate like a role written for him. His relentless investigation of the truth that will destroy him is as much vainglory as

---

*The smith-inventor-artist Daedalus met a more grievous fate than Psyche's for aspiring high. His life has many episodes, of which the most celebrated attributes to him the invention of flight. The designer of the Labyrinth devised wings for himself and his son, Icarus, in order to escape from Crete. In midflight, Icarus fell into the sea after he ventured too close to the sun, whose heat melted his wings. We tend to overlook two essential features of the story in Ovid. Daedalus cautioned Icarus before departure "to fly a middle course." After Daedalus lost his son, the great inventor "cursed his own talents."

courage. "Pride breeds the tyrant" (963) mourns the chorus. And at the end, Œdipus displays no remorse, hardly any sorrow. "What grief can crown this grief? / It's mine alone, my destiny—I am Œdipus" (1495–96). In the other stories, Orpheus and Psyche fail a test and take the consequences. Œdipus' self-absorption in his downfall sounds petulant and childish. But the divinely imposed disasters he has lived through elevate his imperiousness into tragic stature and blindness. We hope desperately that those horrors are his alone, as he proudly affirms. Faust and Frankenstein will aspire to a modified form of this awful greatness.

Do Oriental tales deal differently with these dilemmas of wanting to know more than we should? Not really. The most widely known stories come from *Thousand and One Nights*, a hybrid collection that has entered the mainstream of Western literature. In his justly famous translation-adaptation at the opening of the eighteenth century, the erudite Orientalist Antoine Galland sought out sources and made choices that have affected the Orient's own understanding of that corpus. It is the figure of the genie, or djinn, that concerns us most in its relation to human beings. Distinct from angels, the rough djinns were subdued in Islamic writings into vague gods, similar in most of their behavior to what we would call ghosts. Genies in their infinite guises appear in many Arabian tales as supernatural powers associated with a particular place or object.

On the tenth and eleventh nights, Scheherazade tells the story of a poor fisherman who casts his nets four times and catches only a tightly sealed jar. When he opens it, out rushes an immense genie who fully intends (after telling his story) to kill the fisherman. A little flattery lures the genie to show how he can shrink himself to fit back into the jar. The fisherman claps the lid back on. After several intervening stories, the genie swears by the name of God that he will help the fisherman become rich if he opens the jar again instead of throwing it back into the ocean. The fisherman liberates the genie, and (four stories and sixteen nights later) we learn that the fisherman and his family live out the rest of their days rich and happy. In this case, the evil genie, a Satan figure who rebels against God and against Solomon (Night X), must be kept sealed up in the jar until he has been tamed by so powerful a constraint that he cannot turn his destructiveness upon us. The wily fisherman subdues him by flattering his vanity and then compels him to serve, rather than destroy, human life. No one has given a definitive form to the proverb about keeping the genie in the bottle, but it turns up frequently in our day as a metaphor for controlling science and technology. The fable contains a sequence of events (discovery, fear, outwitting the evil force, prolonged taming, careful release) that resists compression into a formulaic maxim or proverb. We shall consider a less optimistic version of these events when we come to the Frankenstein story.

Carefully considered in their complete versions, the ancient stories of Adam and Eve, of Prometheus and Pandora, of Psyche and Cupid, and even of the genie in the jar appear to give more credence to limits than to liberation, to the dangers of unauthorized knowledge than to its rewards. Ignorance may not be bliss, but the observation of prudent restrictions on knowledge might have prevented the fate of Orpheus, of Icarus, and of Lot's wife.

None of these stories turns entirely on the opposition of knowledge and ignorance. Like Eve and Pandora, Orpheus and Psyche lack faith in the plenitude of their life. They cannot wait. They want more. They come to doubt their own well-being. These two words *faith* and *doubt,* closely shadow any account of knowledge, forbidden and permissible.

It required a poet of epic vision and profound religious devoutness to deal adequately with the motifs of faith and doubt. Banished from the turbulent public life of fourteenth-century Florence and immersed in the theological disputes of the waning Middle Ages, Dante gave in the *Divine Comedy* an imaginary account of himself as an upstart Pilgrim accorded a specially authorized tour of the most restricted zones of Creation. In canto after canto, through Hell and Purgatory and Heaven, the horrors and the marvels that Dante/Pilgrim beholds nudge him toward disbelief. But first Virgil and then Beatrice keep him on the path of faith, and he miraculously survives the lengthy journey through territory forbidden to mortals. The *Divine Comedy* seems to be composed of naïve questions by an outsider who cannot believe what he sees—yet he must believe. Hasn't he seen too much?

In the *Divine Comedy*, the reader and Dante/Pilgrim can never escape from the universe contained within the four opposed words: knowledge or certainty, ignorance, faith, and doubt. In the *Paradiso*, Dante, now guided by Beatrice, has journeyed to the seventh sphere, the heaven of contemplatives, and has come blindingly near his final goal. Peter Damian, a humble sinner who became a reforming cardinal, descends a golden ladder to receive Dante. Feeling himself welcome, Dante makes bold to ask Damian, "Why you alone among your fellow souls / have been predestined for this special task?" This question—is it naïve or unruly?—about the secrets of Providence is cut off short by some disciplinary fireworks, and Peter Damian sends back to Earth through the still-mortal Dante a peremptory message about forbidden knowledge.

The truth you seek to fathom lies so deep
in the abyss of the eternal law,
it is cut off from every creature's sight.

And tell the mortal world when you return
what I told you, so that no man presume
to try to reach a goal as high as this.

(PARADISO, XXI, 94–102, TR. MARK MUSA)

Dante the presumptuous Pilgrim is allowed to proceed on his upward journey, an action that reflects the nascent Renaissance in Italy with its thirst for new knowledge. The rebuke singles out his inopportune curiosity. There are limits on knowledge after all, even after the Poet has been allowed to venture so far.

On the other hand, the very structure of three books, one hundred cantos, and nearly 150,000 verses celebrates Dante's search for knowledge that lies beyond ordinary human knowing. The only slap to his inquisitiveness is administered in the encounter with Peter Damian. Along the way, especially in the final pages of the *Inferno*, Dante includes other incidents that offer a nuanced attitude toward inquisitiveness. Down in the eighth circle of Hell, Dante encounters Ulysses, who has been placed there in punishment for his elaborate deceit of the Wooden Horse to enter Troy. The Pilgrim persuades the Homeric hero to tell how he died, something not supplied to us in the original epic. For the occasion, Dante the Poet invents a whole new tale of further travels for the old warrior-sailor. Too restless to stay home with wife and family, Ulysses and his crew prowl beyond the Pillars of Hercules and cross the Equator, only to meet their death in an immense maelstrom. A little earlier, Ulysses has declared to his crew the impulse behind their endless questing.

> *. . . to this brief waking-time that still is left*
> *unto your senses, you must not deny*
> *experience of that which lies beyond*
> *the sun, and of the world that is unpeopled.*
>
> (INFERNO, XXVI, 114–17, TR. ALLEN MANDELBAUM)

How are we to read this extended digression in which Ulysses occupies more lines than any other personage encountered along the way? Does Dante dream up a whole new ending because, in spite of Ulysses' deceit, the old adventurer's incorrigible restlessness mirrors Dante's?

Before answering, we should look at the incident two books later and one circle deeper in which the great sower of discord, Muhammed himself, displays to Dante his eviscerated body and then issues a sudden challenge. "But who are you who dawdle on this ridge?" Virgil intercedes with a crisp synopsis of the entire enterprise and explains what Dante is doing in the pit of Hell.

> *"Death has not reached him yet," my master answered,*
> *"nor is it guilt that summons him to torment;*
> *but that he may gain full experience,*
> *I, who am dead, must guide him here below. . . ."*
>
> (INFERNO, XXVIII, 46–49, TR. ALLEN MANDELBAUM)

Nothing surprising here—except one word: *esperienza*, "experience." For that was the word Dante used above to designate Ulysses' fatal mission. In Italian, as in French and in Middle English, "experience" refers both to an objective trying-out of something, an experiment, and to the subjective effect of events lived through, the sense of life itself. By the end of the *Divine Comedy*, Dante has implied many times over that he is offering us—piously yet rashly—bootlegged knowledge of things beyond ordinary human ken. Those who perform such a mission, including himself, deserve both admiration and punishment. Once again, one discerns in this deeply medieval author the pull of new learning toward the upheavals of the Renaissance.

Among the four words I have proposed to delimit such enterprises—*knowledge, ignorance, faith,* and *doubt*—Dante interjects a median word, *experience.* This friendly word (which reaches out toward its near homonym, *speranza*, "hope") suggests a secular justification for the presumption that propels explorers like Ulysses and Dante. The appeal to "experience" connects Dante to modern times through Tennyson's poem on the Ulysses theme.

> *Yet all experience is an arch wherethro'*
> *Gleams that untravelled world, whose margin fades*
> *Forever and forever when I move.*

Tennyson's hero could not rest, any more than Dante's Pilgrim could. We have not heard the last of "experience." And in Dante, we can discern a latent and remarkably acute commentary on forbidden knowledge.

Seven centuries after Dante, having lived through the Enlightenment and subsequent revolutions, the West appears to consider itself capable of surviving in a condition of unrestricted knowledge and unbridled imagination. We presume to welcome Prometheus while overlooking Pandora; we do not shrink from looking upon the face of God. I shall pursue these matters in the following sections and come back to them from a different perspective in Chapter VI (on modern science) and in Chapter VII (on pornography and the Marquis de Sade).

2. From Taboo to Science

In the early sixteenth century, two historical developments converged, with far-reaching effects in Europe. The impulse to reform the Catholic church led to the writings and translations of Martin Luther and to the formation of the first heretical Protestant sects in Germany. And the spread of movable type made possible the printing and distribution of books on an unprecedented scale. Even the prospect that ordinary people might read the Bible for themselves—let alone works of modern heresy—challenged the authority of the Church. These were the circumstances that led to the institution in 1559 of the *Index Librorum Prohibitorum*. The condemning and burning of books had occurred spasmodically during the Middle Ages. Universities sometimes drew up their own lists. Now the Church itself sought to control what came off the printing presses and what could be read.

In the face of Enlightenment ideas about freedom of speech and religion, the *Index* has not been a successful device for defending the Church against its enemies. But we should not mock it too unthinkingly. Like Plato banishing poets from his Republic, an index could be interpreted as attributing more efficacy, more significance, and therefore more potential risk to ideas and words than does a policy of unrestricted free speech. Tolerance belittles. Exiles from repressive regimes often observe that freedom trivializes courageous thinking. Furthermore, the *Index* did not usually destroy works; it restricted access to forbidden books to scholars—surely not the most obedient readership.

But the West has turned in a different direction. Gradually, we have replaced the *Index* and other forms of censorship with the free marketplace of ideas and a liberal education. And we have almost forgotten how bold a social experiment we have undertaken and how much devotion will be required to make it work.

Out of the same sixteenth century that codified the *Index* came Michel de Montaigne, fearlessly frank in discussing ideas and in describing the foibles of his own personality. Companion of kings, and later mayor of Bordeaux, he returned at forty to a world of books and writing in the tower study of his château. Through his continually expanded *Essays*, the supple form of writing he invented, we probably know more about Montaigne's inner life and tastes than of any important historic figure, including St. Augustine and Rousseau. Above all, he despised people who puffed themselves up. The lengthy essay he wrote after reaching forty, "Apology for Raymond Sebond," though often playful and relaxed, shows little sympathy toward human aspirations. "Presumption is our natural and original malady. . . . it is by the vanity of this very imagination that man sets himself up as the equal of God." Human presumption to knowledge would provide the title of a later essay (Book II, Chapter 17) and a recurring motif carried through to the pages of the last great essay, "Of Experience" (Book III, Chapter 13). In the "Apology," Montaigne on the perils of the imagination echoes *Genesis* and likens it to curiosity. "Christians have a special knowledge of the degree to which curiosity is a natural and original evil *[mal]*." The vocabulary reveals the depth and clarity of Montaigne's conviction about the presumptuous curiosity of our imagination. Little wonder that after referring to the temptation of Eve and Adam, and of Ulysses "offered the gift of knowledge by the Sirens," Montaigne sounds like the enemy of philosophy. "That's why ignorance is so strongly recommended to us by our religion as the appropriate path to belief and obedience." After many pages demonstrating the weakness of our senses and our judgment, Montaigne concludes, like Socrates, that ignorance aware of itself is the only true knowledge. Ten years later, his last essay contains sentences that show he has yielded no ground. "Oh what a sweet soft pillow ignorance and incuriousness provide for a well-made head."

Like most creatures of the mind, Montaigne could not follow the advice he formulated out of his own experience. As with most of the figures in this book, his inquisitiveness knew no bounds. The contradiction should not surprise us. The anti-intellectualism of an intellectual (there were no such terms in the sixteenth-century) probably qualifies as presumption to the second power.

In matters of religion, Montaigne accepted the Catholic faith not on the basis of reason but by ironic conformity to traditional beliefs. The position disturbed no one very much during Montaigne's lifetime. By the middle of the next century, however, Catholic theology had become rationalist enough to assign him a place on the *Index* because of his fideism (overreliance on faith alone) and his distrust of human faculties. Despite his deeply held skepticism about the powers of reason, Montaigne never stopped reading and thinking and writing. The frankness with which he dealt with these contradictions speaks directly to us today.

Montaigne's reluctant disciple in the seventeenth century, Blaise Pascal, had many claims to fame: mathematician (Pascal's law), inventor of the roulette wheel, hair-shirt mystic, powerful religious pamphleteer, and, in the fragmentary *Pensées*, incomparable psychologist, Pascal shared Montaigne's wariness of the imagination. Their common attitude is borne out by the metaphor they both picked to describe our conduct in the field of knowledge. The concrete word they found was *portée*—the reach of an arm, the range of a weapon, the significance of an event or idea. Our "reach" defines both our capacity and our limit, complementary aspects of our character. We have to know both and distinguish them. "A man can be only what he is and can imagine only according to his reach *[portée]*," writes Montaigne in the "Apology" (501). Those two instances of "can" might well be read as "should." At the end of the same essay, Montaigne makes clear that *portée*, the appropriate scale in all things, contains the remedy for presumption. "To make a fistful bigger than our fist, an armful bigger than our arm, to hope to step further than the length of our legs—these actions are impossible and monstrous. The same goes for man's attempt to rise above himself and humanity" (588).* Pascal had read Montaigne attentively and, in his magnificent *pensée* on the two infinities, adds dignity to the metaphor. "Let us then know our reach *[portée]*. We are something, and not everything. . . . Our intelligence occupies in the order of intelligible things the same place as our body in the extent of nature" (Lafuma number 199). Montaigne's philosophical skepticism about our faculties of curiosity and imagination, about our incorrigible vanity and presumption, produces the final and most graphic image of the *Essays*. "On the highest throne in the world we can sit only on our own arse *[cul]*" (1096).

In speaking of the human itch to overreach, Montaigne and Pascal remained fairly lighthearted. In prehistoric and primitive societies, similar concerns about forms of forbidden knowledge have been dealt with under a more ominous term: *taboo*. The word is

*One wonders how Montaigne could have missed the earthy French proverb: *"N'essaie pas de péter plus haut que ton cul."*

Polynesian; a useful definition comes from Frazer's *The Golden Bough.* Taboo refers to an object, place, person, or action in which "holiness and pollution are not yet differentiated." In *Totem and Taboo,* Freud closely follows Frazer and describes a fusion of sacred and forbidden. Frazer's and Freud's lengthy enumerations of taboos in primitive societies emphasize two complementary aspects.

> *To* [the savage] *the common feature of all these* [tabooed] *persons is that they are dangerous and in danger, and the danger in which they stand and to which they expose others is what we should call spiritual or ghostly, and therefore imaginary. The danger, however, is not less real because it is imaginary.*
>
> (*THE GOLDEN BOUGH,* Chapter XXI)

> *Taboos are very ancient prohibitions which at one time were forced upon a generation of primitive people from without, that is, they probably were forcibly impressed upon them by an earlier generation. These prohibitions concerned actions for which there existed a strong desire.*
>
> (*TOTEM AND TABOO,* Chapter II)

Every myth, every tale I have mentioned, deals with an awakening to the dilemma of curiosity about something both attractive and dangerous. Freud used the word *uncanny (das Unheimliche)* to cover some of the same territory. All these terms testify to the protective principle that Frazer compares to the operation of "electric insulators." The force of taboo insulates "the spiritual force" in the object or person from violation and also insulates us—at times not adequately—from its forbidding yet alluring power.

Montaigne's and Pascal's warnings against curiosity and presumption, their down-to-earth version of taboo, occurred just at the moment when the great creature we now call "science" was beginning to stir. How then did the nonreligious disrupting force of science gain admission into a culture based primarily on custom and on faith? To answer, I shall back up a little bit in order to approach a key seventeenth-century figure not yet mentioned.

Up through the Middle Ages, Christian theology incorporated and imposed upon the faithful a dark suspicion of secular nature.

Our proper devotion should be to the divine order of grace. St. Paul and St. Augustine warn us continually to distrust the original curiosity of Adam and Eve in a Satan-haunted world. The literary scholar Basil Willey observes that well into the seventeenth century, secular knowledge and natural philosophy represented "a distraction or seducement" from true spiritual living. "To study nature meant to repeat the sin of Adam." Nevertheless, like a slow-moving glacier, Christian theology trundled along within it some unassimilable boulders. In 1336, Petrarch, celebrated for his love poetry in Italian, climbed Mount Ventoux in Provence just "to see what so great an elevation had to offer." He said he almost lost his soul at the summit "admiring earthly things," like the view. Yet later, he wrote an astonishing letter to record the pleasures of that excursion into nature. Petrarch came to value the secular world as highly as Dante valued the spiritual.

Petrarch's secularizing inclinations can be traced during the fifteenth and sixteenth centuries through the witty satires of the Dutch humanist Erasmus and the un-Christian statecraft of Machiavelli down to the exemplary career of Galileo in the seventeenth century. For him, authority did not lie in Aristotle or in Genesis or in Christian theology. It lay in the revelations of the device he patiently ground and mounted in order to behold the heavens thirty times larger than with the naked eye. He could then demonstrate what the Pole Copernicus had only hypothesized about where the center of our own system lies. Galileo, however, inevitably ran into trouble with ecclesiastical authorities, who placed him under house arrest. His investigations could not be assimilated by a Christian society. They remained forbidden. In Italy, science was at an impasse. It required a change of scene.

The figure on whom this account of expanding knowledge now pivots is a scalawag English statesman of stunning intellect. He helped prosecute his own protector and take him to the scaffold. In 1621, at the peak of his career as lord chancellor, he was brought low by a bribery scandal and served a short prison term. Francis Bacon had published his first pithy *Essays* in 1597, when he was thirty-six. He knew Montaigne's *Essays* well and modified both the form and the message to suit his own purposes. Eight years later, after James I had promoted and knighted him, Bacon wrote *The Advancement of Learning*, a book that shifted the orientation of intellectual endeavor in his time and helped open the way for the Enlightenment. What he also called *The Great Instauration* undertook to convert Aristotle's deductive logic into inductive investigation. One could not yet call it scientific method. In question-and-answer argument echoing St. Thomas Aquinas, Bacon quotes Scripture to right and left (especially Ecclesiastes) in order to demonstrate that "God has framed the mind like a glass, capable of the image of the universe. . . . Let no one weakly imagine that man can search too far, or be too well studied in the book of God's word, and works, divinity, and philosophy" (Book I). Only the desire for "proud" knowledge of good and evil betrays our humanity and rivals God. The "pure" knowledge of nature contemplates and glorifies God's works. Thus Bacon refutes the argument that the pursuit of knowledge "hath something of the Serpent and puffeth up." He was astute enough to leave higher theology to the theologians. We could call him the Great Compromiser.

The Advancement of Learning made a timely and powerful argument in favor of science as belonging to God, not to the Devil. In the unfinished utopia, *The New Atlantis* (1627), the careful pages Bacon devoted to Salamon's House describe it as a semiecclesiastical scientific-research institute whose activities represent a form of worship and giving "thanks to God for his marvellous works." Bacon himself made no significant scientific discoveries. But his championing of scientific research facilitated the landmark work in the seventeenth century of the physiologist William Harvey, who demonstrated that blood circulates in the body by the pumping action of the heart, and of Robert Boyle, who established the nature of chemical reaction. Bacon's ideas led to the founding of the Royal Society after his death. In proclaiming that the new world of geographic exploration and scientific discovery required a new philosophy, Bacon displayed, as Basil Willey writes, a "magnificent arrogance" in his political career, in his scientific attitude, and in his varied and apposite prose—both in Latin and English. He claimed all philosophical knowledge for his domain and also identified "the deepest fallacies of the human mind" in terms that have become proverbial: the idols of the tribe, of the den, of the marketplace, of the theater. For Bacon, the prophet of modern science and its earliest poet, true scripture lay in the infinite book of nature, as it did for Galileo. Bacon broke the taboo against science. After his assault on the idea of forbidden knowledge, scientific endeavor and the accompanying doctrine of progress have for four centuries encountered fewer obstacles.

In a lengthy career that combined "civil business" (i.e., politics) and philosophy, Bacon became very aware of different forms of knowledge. Several of his writings adapt Montaigne's *Essays* and distinguish three types of philosopher: those who think they know the truth, or presumptuous dogmatists; those who believe nothing can be known, or despairing skeptics; and those who keep asking questions in order to extend imperfect knowledge, or persistent inquirers. Favoring the last intermediate category as pointing to the future of philosophy, Bacon also associated it with the pre-Socratics.* Still a believer, he did not abandon all limits in his liberation of science. These three categories—presumptuous dogmatists, systematic skeptics, and persistent inquirers—have not lost their pertinence in a discussion of forms of knowledge, forbidden and otherwise.

3. SKEPTICISM, AGNOSTICISM, *IGNORABIMUS*

Following Bacon's repeated insistence on *induction* in the pursuit of truth, it is possible to trace an essential Enlightenment tradition leading to modern technological and scientific achievements. At the close of the twentieth century, we speak confidently of our research institutes and our institutions of higher learning as of officially sanctioned enterprises opening up enhanced vistas of life through the conquest of nature. We are intent on cracking the secrets of the atom and the genetic code as well as those of outer space. To track it all we have an "information superhighway." On the other hand, the term *knowledge explosion* expresses our anxiety about the potentially devastating consequences of such research. We are not at ease with our new Temple of Solomon or Tower of Babel. The strand of my story that I shall follow down to the present in the

*Kenneth Alan Hovey has published a fine article on the evolution of Bacon's thought on these questions and its relation to Montaigne's.

remainder of this chapter never strays far from the question of the limits of science.

A copious store of proverbs and parables cautions us about the presumptions and delusions of learning. Bacon's own distinction between *pure* and *proud* learning leads to his warning against "confounding the two different streams of philosophy and revelation together." When he finally reaches Book IX of *The Advancement of Learning,* he ostentatiously omits theology as something issuing not from science but from the word and oracles of God. Bacon's great plea for secular knowledge and systematic research ends with a prayer "to the Immortal Being through his Son, our Saviour."

The careful balance of intellectual courage, respect of religion, and political expedience in Bacon's work survived virtually unshaken for over a century and reemerged in Pope's early writings. The famous couplet that opens Epistle II of *An Essay on Man* (1734) epitomizes both Pope and Bacon. Presumption comes back like an old refrain.

> *Know then thyself, presume not God to scan;*
> *The proper study of Mankind is Man.*

Voltaire achieves a comparable terseness of expression. When, after a lifetime witnessing the sufferings and duplicities of mankind, Candide can at last profess that he knows something, he quietly sets aside all Pangloss' claims to metaphysical knowledge and makes a modest proposal: "let us cultivate our own garden." In *Candide,* Voltaire produced a sassy parable on the theme of *portée,* of living within our reach or range, a theme he inherited from Montaigne and Pascal.

In his satirical fiction, Swift approached the problem of knowledge in an equally concrete fashion. Gulliver describes the wonderful invention of gunpowder and cannon to the King of Brobdingnag; the King is "struck with horror" and protests "that he would rather lose half his kingdom than be privy to such a secret, which he commanded me, as I value my life, never to mention any more" (Book II, 7). Few minds ranged as freely as Voltaire's and Swift's across the landscape of knowledge, religious and secular. And few minds became so preoccupied with the errors and dangers of "proud learning" in all human endeavor.

In cautioning us through their satirical fiction against overconfidence in reason, Voltaire and Swift were not referring only to destructive technologies of war. They distrusted the tendency of high intellect to seek rarified speculations and empty categories. Twice, in chapters V and XXI of *Candide,* Voltaire interrupts a discussion of free will with an ellipsis in midsentence, as if to say that we waste our time trying to solve ultimate metaphysical questions. In the third book of *Gulliver's Travels,* Swift portrays the mathematically brilliant and ambitious Laputans, whom Gulliver discovers living in the clouds. The Laputans are characterized principally by having "one of their eyes turned inward, and the other directly up to the zenith" (III, 2). Stumbling often, they have discovered neither any ultimate truth nor a modest garden to cultivate.

One of the most comprehensive and arresting statements affirming the path of reason comes from Thomas Jefferson writing about founding the University of Virginia. It was the first secular university in a new nation without an established church. Jefferson's Enlightenment optimism has shaken off any lingering sense of knowledge as the work of the Devil. He wrote: "This institution will be based on the illimitable freedom of the human mind. For here we are not afraid to follow the truth wherever it may lead, nor to tolerate any error so long as reason is left to combat it" (December 27, 1820: to William Roscoe). It sounds as if Jefferson were writing while looking at the opening page of Kant's 1784 essay, "What is Enlightenment?" For in effect, Jefferson reaffirms the motto from Horace that Kant quotes in his opening paragraph: *Sapere aude,* "Dare to know!" Jefferson ignores the expedient social and political constraints tacked on by Kant. It also sounds as if Jefferson were following Jesus' adjuration to the Pharisees: "Ye shall know the truth and the truth shall make you free" (John 8: 32). But Jesus' truth is revealed and eternal rather than a secular knowledge discovered by our own investigations.

Jefferson, founding not a republic but an institution of higher learning, produced a declaration of rationalism unsurpassed in American and European intellectual history. However, this sturdy rationalism had to accommodate itself to a lingering strain of restraint and skepticism about science that the best scientists would not conceal even in the face of a comprehensive new theory of evolution. During the turbulent decade that followed the publication of *The Origin of Species* (1859), Darwin found his stoutest champion in Thomas Henry Huxley, a young biologist educated on Carlyle, Goethe, and Schelling and trained to science (like Darwin himself) during a four-year naturalist's voyage to the Pacific. In 1860, Huxley was a thirty-five-year-old professor at the School of Mines. At a packed meeting in Oxford of the Zoological Section of the British Association, Huxley listened quietly to Bishop Wilberforce's famous mocking question: "I should like to ask Professor Huxley . . . if it is on his grandfather's or his grandmother's side that the ape ancestry comes in?" The fact that in the mid-nineteenth century many educated people were losing their Christian beliefs and their faith in the literal truth of the Bible led their opponents to counterattack. Rising to respond, the tall, stern Huxley first gave a lucid summary of Darwin's ideas on natural selection and then proceeded with relish to the question of ancestry.

> "*. . . a man has no reason to be ashamed of having an ape for his grandfather. If there were an ancestor whom I should feel shame in recalling it would rather be a* man—*a man of restless and versatile intellect—who, not content with an equivocal success in his own sphere of activity, plunges into scientific questions with which he has no real acquaintance, only to obscure them by an aimless rhetoric, and distract the attention of his hearers from the real point at issue by eloquent digressions and skilled appeals to religious prejudice.*"*
>
> (LIFE AND LETTERS, I, 199)

Huxley's adroitness in turning the fire back on the attacker came out again at an early meeting of another association of clerics, scholars, and men of science. One member urged the need to avoid in the debates any "moral disapprobation of fellow members" and to shun personal attacks. W. G. Ward, an Anglican cleric recently converted by Cardinal Newman to Roman Catholicism, demurred. "While acquiescing in this condition as a general rule, I think it

*This is the account given by John R. Green, then an undergraduate at Oxford. Thirty years later, Huxley said that Green's account was substantially correct, except that he was certain he had not used the word *equivocal*.

cannot be expected that Christian thinkers shall give no sign of the horror with which they would view the spread of such extreme opinions as those advocated by Mr. Huxley." Ward was echoing the matron whose remark I quote in my foreword. All accounts report a brief pause followed by this reply from Huxley: "As Dr. Ward has spoken, I must in fairness say that it will be very difficult for me to conceal my feeling as to the intellectual degradation which would come of the general acceptance of such views as Dr. Ward holds."

This exchange took place at the Metaphysical Society, organized in 1869 by the broad-minded editor and intellectual impresario James Knowles, seconded by Tennyson, the poet laureate. Ten years after Darwin's *Origin*, debate still raged at such a pitch that some spoke seriously of a New Reformation, and others felt that civilization itself was crumbling before atheism and nihilism. Knowles persuaded all parties to join the discussions of the Metaphysical Society, from Archbishop Manning to Roden Noel, "an actual atheist and a red republican." Among the great English minds of the day, only J. S. Mill, Cardinal Newman, and Herbert Spencer refused the opportunity to air their ideas. The bishop of Peterborough declared, "We wanted only a Jew and a Mohammatan" among the sixty-odd members to complete representation of all faiths.

During the organizational meetings of the Metaphysical Society, Huxley became very impatient with the compulsion put on him to accept a label for his philosophical position.

When I reached intellectual maturity, and began to ask myself whether I was an atheist, a theist, or a pantheist; a materialist or an idealist; a Christian or a freethinker; I found that the more I learned and reflected, the less ready was the answer; until, at last, I came to the conclusion that I had neither art nor part with any of these denominations, except the last. The one thing in which most of these good people were agreed was the one thing in which I differed from them. They were quite sure they had attained a certain "gnosis"—had, more or less successfully, solved the problem of existence; while I was quite sure I had not, and had a pretty strong conviction that the problem was insoluble. And, with Hume and Kant on my side, I could not think myself presumptuous in holding fast by that opinion.

(LIFE AND LETTERS, I, 343)

Huxley had, in other words, a strong philosophical position, for which there was no accepted name. He was too resourceful to be plagued for long by this problem of nomenclature. Having decided that he was being treated like a "fox without a tail," he made a brilliant strategic move by using the English language as his field of maneuver.

*So I took thought, and invented what I conceived to be the appropriate title of "agnostic." It came into my head as suggestively antithetic to the "gnostic" of Church history, who professed to know so much about the very things of which I was ignorant; and I took the earliest opportunity of parading it at our Society, to show that I, too, had a tail, like the other foxes. To my great satisfaction the term took.**

(LIFE AND LETTERS, I, 343–44)

It is hard to believe that Western languages survived until the mid-nineteenth century without an equivalent of *agnostic.* Yet Huxley was not mistaken. Words like *freethinker*, *libre penseur*, *libertin*, *deist*, *theist*, *atheist*, and *heretic* all referred to holding positive convictions on large metaphysical questions. *Skeptic* and *Pyrrhonist* connoted systematic doubt in all domains. Such terms carried connotations far removed from Huxley's uncertainty about final questions and his certainty about "natural history" or science. There was virtually an empty space in the language, like a gap in the periodic table awaiting the discovery of a new chemical element.

At exactly the same period, Darwin apparently experienced a comparable need to define his philosophical position. His letter to J. D. Hooker in 1870 is perfectly frank. "My theology is a simple

---

*Huxley's "term" probably alludes also to St. Paul's mention of an altar "To the Unknown God" (Acts 17:23).

muddle; I cannot look at the universe as the result of blind chance, yet I can see no evidence of beneficent design, or indeed of design of any kind, in the details." Six years later in his *Autobiography*, Darwin first calls himself a "Theist" and then goes a step further. "The mystery of all things is insoluble by us; and I for one must be content to remain an Agnostic." The word could now be considered certified. It represents a modest yet unflinching form of forbidden knowledge.

Huxley's neologism did not come to him entirely by momentary inspiration. Years before, in September 1860, he had been obliged to think through his religious and scientific views when his healthy four-year-old son died suddenly of scarlet fever. In a letter of condolence, Charles Kingsley, author of *Westward Ho!* and chaplain to Queen Victoria, tried to comfort him with the doctrine of immortality. Huxley responded with a ten-page letter that reveals anguished feeling and unshakable intellectual integrity. In those pages, he elaborated the agnostic position without using the word. "I neither deny nor affirm the immortality of man. I see no reason for believing it, but, on the other hand, I have no means of disproving it." The letter remains clear-sighted and honest in the midst of deep personal distress.

In 1889, twenty years after the founding of the Metaphysical Society, Huxley found himself drawn into a new controversy, which surrounded the tedious, hugely popular three-volume novel *Robert Elsmere* by Mrs. Humphry Ward, Matthew Arnold's niece. The novel contained a sustained attack on biblical miracles as lacking adequate testimony to compel belief. In mounting their counterattack against this kind of thinking, spokesmen for the Church saw advancing against them on all sides a new enemy, who, in their descriptions, sounded like Huxley. "He may prefer to call himself an agnostic; but his real name is an older one—he is an infidel." Huxley joined the fray with four new articles on agnosticism and even cited in support of his position Cardinal Newman's ideas about the evolution of the Catholic church. At bottom, Newman was probably more beset by doubts about mankind's role in the world than Huxley, who kept faith with "the wild living intellect of man" and believed in the future. But Huxley's new word became deeply enmeshed in the religious controversies of his day.

The aspect of these historic debates that most concerns us is the two meanings of the word *agnostic* that emerge from Huxley's writings, and that still hover around the term. The letter to Kingsley and the statements to the Metaphysical Society in 1869 give to *agnostic* this categorical sense: The human mind alone cannot answer the ultimate questions of metaphysics and theology and cannot know "true" reality behind appearances. These things are beyond us. On the other hand, Huxley's later writings attach agnosticism to a larger and older tradition that links the method of Socrates with the Reformation and with Descartes. "In matters of intellect, follow your reason as far as it will take you." Accept nothing without demonstration. You might even reach ultimate truth. Huxley's modified attitude takes the emphasis off the notion of limits and attaches the word to a gentle skepticism, almost to what we now call pragmatism. Several of Huxley's contemporaries believed that he had compromised a useful word by softening its meaning.

For my purposes in writing about forbidden knowledge, the first, rigorous meaning evidently carries the greater intellectual weight and should be the primary meaning attributed to the word. *Agnostic* refers not only to recognizing our ignorance about ultimate questions but also to the claim that those problems are "insoluble," as both Darwin and Huxley wrote, beyond our reach. This articulate, argumentative biologist who attacked the certainty, the *gnosis*, of others while restraining his own did not propose to stop the march of either science or religion. He did, however, challenge his generation to scrutinize soberly the claims made by both camps and gave us a new word as a talisman of unobtrusive doubt.*

Three years after Huxley's astonishingly successful coinage of a term for his philosophical and religious position, a German scientist, twice rector of the University of Berlin, delivered a celebrated lecture, entitled "On the Limits of Science." Emil Du Bois-Reymond (1818–1896) had acquired an extensive knowledge of French intellectual culture in addition to his German scientific training. His careful laboratory research on electric fish and his development of

---

**The Oxford English Dictionary* accurately records the origin of *agnostic* that I have just outlined. A related term coined a few decades earlier belongs as much to psychology as to philosophy. In 1830, Auguste Comte proposed *altruism* to designate a principle of conduct based on the interests of others and opposed to egoism. *Altruism* refers to the optimistic strain in Enlightenment thinking and loosely parallels the connotations of philanthropy and benevolence.

experimental apparatus such as mercury switches and current multipliers had earned him wide respect in the field of physiology. Later in his career, he lectured widely on the scientific significance of such writers as Voltaire, La Mettrie, Diderot, and Goethe.

By the time of his lecture in 1872, Du Bois-Reymond had already made a reputation among scientists and intellectuals as a strong opponent of "cosmic consciousness," a popular notion substituting for deity or divine mind. At the opening of the nineteenth century, Laplace had said that in searching the heavens with a telescope he found no God. Du Bois-Reymond made a comparable materialist refutation of cosmic consciousness, affirming that he found no evidence anywhere of cosmic neural tissue fed by arterial blood and "proportional in size to the faculties of such a mind." He had also attacked the hypothesis of a "vital force," calling it an appeal to the supernatural in order to account for the step from inorganic to organic. Consequently, Du Bois-Reymond had powerful credentials as a no-nonsense scientist who believed in natural causation, not in metaphysical entities. Because of those credentials, the 1872 lecture on the limits of science shocked many of his colleagues.

In a change of heart since his early research on animal electricity, Du Bois-Reymond now affirmed that he saw serious gaps in the explanatory power of materialist science. He revived the category of what medieval philosophers had called *insolubilia*—problems beyond solution by science, such as "Why is there anything at all?" Du Bois-Reymond ended his lecture with the Latin term *ignorabimus*—"we shall remain ignorant." In a later lecture, "The Seven Riddles of the Universe" (1880), he proposed that at least three of the great foundational issues in physics, biology, and psychology transcend man's scientific capacities.* Toward the close of a century that prided itself on scientific prowess, these were fighting words. To speak of "limits" on science sounded like a stronger

*His seven riddles retain a certain pertinence: the existence and nature of matter and force; the origin of motion; the origin of life; the nature of adaptation in organisms; the origin of sensory perception; the origin of thought and consciousness; and the problem of free will. Some of them might be solvable in the future, but not all. Du Bois-Reymond did not clarify the relative degree to which the *insolubilia* owe their status to the nature of the universe or to the nature of the inquiring human mind.

version of Bacon's distinction between "proud science" and "pure science." The glossy Latin label *ignorabimus* refused to recognize some higher certainties of science, and made that refusal as steadfastly as did the term *agnosticism.* Both terms outraged alike convinced believers and convinced unbelievers.

The most stentorian response to Du Bois-Reymond came from the zoologist Ernst Haeckel. This polemecist and popularizer of science was the chief defender on the Continent of Darwin's ideas and of the biogenetic law that ontogeny recapitulates phylogeny. (Today, Haeckel's law has regained considerable prestige.) His book *The Riddle of the Universe (Die Welträtsel)*, published in 1889, caught the imagination of many readers and appeared in several editions. It was the era of spiritualism and Theosophy. After dismissing free will as a pseudoproblem that "rests on mere illusion and in reality does not exist at all," Haeckel claimed that all the other riddles except one had been solved. The remaining "problem of substance," the ultimate origin of matter and its laws, he regarded as more metaphysical than scientific.

A clear response to this nineteenth-century controversy comes in the work of a contemporary American scholar, Nicholas Rescher. This indefatigable historian and philosopher of science has preceded me over some of the terrain explored in this book. In *The Limits of Science* (1984), Rescher condemns both parties in the controversy. Du Bois-Reymond had only the shakiest of grounds on which to extrapolate the existence of *insolubilia* out of our present ignorance; Haeckel was equally wrong in implying that science at the end of the nineteenth century was approaching completion of its tasks and that soon all the answers would be known. Rescher easily demonstrates that "the perceived completeness of science" is an illusion. Nature is inexhaustible; it has no bottom. Our questions never cease: There are no final truths.* In rejecting any form of omniscience in science as well as Du Bois-Reymond's *insolubilia,* Rescher in effect proposes a modified form of *ignorabimus* related to agnosticism. For he posits an unending procession of questions addressed by our cognitive faculties to the apparently bottomless

---

*Rescher wrote just a few years too early to deal with the debates in the early nineties over the claims of a "final theory" made by advocates of the superconducting supercollider and of the Higgs particle.

well of nature. We can never exhaustively know a world system that contains us as well as our ongoing investigations into that system and our modifications of it.

Rescher does not stop there. He goes on to write an essay that looks beyond practical considerations to deal with the moral, prudential, and legal limits we might wish to apply to scientific inquiry. In the process, he looks far beyond science. He wonders briefly what would happen if we could devise some means of seeing into other people's minds to know their intentions and motives. His conclusion startles us by its wariness.

> *Some information is simply not safe for us—not because there is something wrong with its possession in the abstract, but because it is the sort of thing we humans are not well suited to cope with. There are various things we simply ought not to know. If we did not have to live our lives amidst a fog of uncertainty about a whole range of matters that are actually of fundamental interest and importance to us, it would no longer be a human mode of existence that we would live. Instead we would become a being of another sort, perhaps angelic, perhaps machine-like, but certainly not human.*
>
> *There is a more deeply problematic issue, however. Are there also moral limits to the* possession *of information* per se*—are there things we ought not to know on moral grounds? . . . Here, inappropriateness lies only in the mode of acquisition or in the prospect of misuse. With information, possession in and of itself—independently of the matter of its acquisition and utilization—cannot involve moral impropriety.*
>
> ("FORBIDDEN KNOWLEDGE," 9)

Few authors have faced these questions so directly and unshrinkingly as Rescher. But he develops them no further and returns to his central concern with science. Therefore, I quote his two paragraphs as a takeoff point for my own investigation, which will turn to literature and come back to science much later.

I disagree with Rescher on only one point. The second, presumably scientific issue of possession of knowledge (versus its acquisition and use) does not impress me as any more "deeply problematic" than the first issue, how our humanity may be revealed and even defined by "liv[ing] our lives amidst a fog of uncertainty about a whole range of matters that are actually of fundamental interest and importance to us." These words, which echo phrases like St. John of the Cross's "cloud of unknowing" and Keats' "negative capability," locate a certain ignorance at the very seat of human nature. That fog turns out to be our nimbus and our protective veil. We shall have to return many times to these paradoxes as we enter further into the subject.

To this remarkable short essay, Rescher gives the title "Forbidden Knowledge: Moral Limits of Scientific Research." He goes on to say that "it is the basically correct moral of [the Garden of Eden] story that we may well have to pay a price for knowledge in terms of moral compromise." This judicious philosopher of science makes bold to add explicitly human and moral dimensions to the technical terms *insolubilia*, *agnostic*, and *ignorabimus*. The exploration of these dimensions will carry us far beyond science.

4. "LUST OF THE SOUL"

The famous scientists mentioned on page 13 all said they were motivated in their work by curiosity above all. A large collection of myths and stories, from Pandora's opening a jar to Petrarch's climbing a mountain, suggests we cannot escape curiosity. But is the picture that clear? Other evidence may give us pause.

The Wild Boy of Aveyron came out of the woods in 1800, when he was aged twelve, behaving like an animal after several years of isolation. Dr. Itard, who observed and trained him for the next dozen years, made much of the fact that the boy had to be taught to imitate, since he had no natural instinct or inclination to copy what he saw others doing around him. The thick detail of Itard's reports and of many parallel reports also describes an even more fundamental condition: The Wild Boy felt no curiosity. Beyond his need for food and sleep, he sought nothing else. He was content to rock on his haunches and to vegetate wherever he was. He seemed immune to the curse of boredom. Without being conclusive, the case offers glimpses of inchoate human nature.*

Is curiosity—the desire to know more than is necessary for our immediately foreseeable needs—acquired in childhood, or is it given to us in some inherited form? *Ignorabimus*. In either case, our curiosity has become self-conscious and self-sustaining. Possibly a full and final answer about the origins of curiosity is one of the things we should *not* know if we are to remain human, if we are to keep the fog of uncertainty that defines us. But such a response troubles those of us who have been brought up to believe in Jefferson's pursuit of truth wherever it may lead. We are reluctant to connect that attitude with hubris and presumption and to acknowledge Montaigne's and Pascal's appeal to *portée*, our ordinary reach.

The fragments of a history of forbidden knowledge that I have outlined in this chapter lead forward toward significant works and episodes that will take us far deeper into the subject. The slender outline already sketched points to a certain fluctuation within a steady state of affairs, to a dynamic equilibrium between a presumptuous pursuit of knowledge and a skeptical, cautious approach to it. Even the persons and the periods most confident of the virtues of knowledge—Plato, say, and the Enlightenment—contain their own powerful compensating mechanisms. Socrates knew best that he did not know. No one has mocked the abuses of reason more effectively than Swift and Voltaire, who represent the Age of Reason. We have not advanced beyond the interlocking notions of liberation and limits.

And such a history also demonstrates that popular wisdom residing in proverbs and legends does not lie far away from the intellectual scruples affirmed in more recently minted terms such as *agnostic* and *ignorabimus*. "Curiosity killed the cat." "Let sleeping dogs lie." But what kind of a paradox is this? Must I cease and desist from the very inquiry that beckons me most? Should I be ashamed of my curiosity? We seem to be dealing with a convergence of opposites in ourselves, a mental condition analogous to the bodily condition W. B. Yeats describes as vividly as any proverb could.

*See my *The Forbidden Experiment: The Story of the Wild Boy of Aveyron* (1980).

But Love has pitched his mansion in
The place of excrement.

("CRAZY JANE TALKS TO THE BISHOP")

Our yearning for knowledge was long ago dubbed *libido sciendi*, a term that insists on the analogy between curiosity and sexual desire. In Book X of the *Confessions*, in which St. Augustine describes our three major temptations, he closely associates "concupiscence of the flesh," particularly sexual lust, with concupiscence of the eyes. He means lust for knowledge, which is "in many ways more dangerous."

> *There is also present in the soul, by means of these bodily senses, a kind of empty longing and curiosity, which aims not at taking pleasure in the flesh but at acquiring experience through the flesh, and this empty curiosity is dignified by the names of learning and science. Since this is in the appetite for knowing, and since the eyes are the chief of our senses for acquiring knowledge, it is called in the divine language* the lust of the eyes.
>
> (CHAPTER 35)

Self-confessed sinner and Christian convert, St. Augustine insists on the concupiscence of the mind as more perilous than that of the flesh. And like Dante and Tennyson, he finds the word *experience* to designate the object of this "empty curiosity." Can the dignified words *learning* and *science* be swept aside so peremptorily? No, but St. Augustine's insight into the dynamics of human knowing also stands, animated by the word *concupiscence*.

Thomas Hobbes made a similar association in the seventeenth century. In Chapter Six of *Leviathan*, where he is still calling a preliminary role of human emotions, natural lust and luxury are followed by curiosity: "Desire to know how and why, CURIOSITY . . . is a lust of the mind, that by a perseverance of delight in the continual and indefatigable generation of knowledge, exceedeth the short vehemence of carnal pleasure."

Yet, St. Augustine and Hobbes wrote these severe words in the midst of a project to discover and report knowledge about the how and why of human actions. The desire to know the world and other beings as well as ourselves belongs both to our highest aspirations, celebrated by Homer and Dante and all great authors, and to our basest concupiscence in wanting to reach beyond our *portée*. Because it feeds both our glory and our shame, curiosity provides the motif of many of our greatest narratives of quest and conquest, of love and passion.

Chapter II

MILTON IN THE GARDEN OF EDEN

■

1. Resistance to Adam and Eve

From Pandora, the first woman sent to tempt mankind, to a meeting of scientists discussing the origins of their vocation, curiosity claims a major role in our lives. At the same time, we have to register limits to knowledge, limits lodged in our minds and impediments intrinsic to the nature of the universe itself. No story records these conflicting motifs more simply and convincingly than the human segment of the earliest Hebrew creation myth, which opens the first book of the Torah.

These opening episodes offer us answers to three ancient and troubling questions. How did everything begin? Why does life bring so much suffering, deceit, and destruction—so much positive evil? Why do we die? Thus we have the question of origins, the question of theodicy, and the question of mortality. The stories of the creation and of Adam and Eve that open Genesis accept the three challenges these questions represent. A close reading shows that the chapters combine at least two independent sources, which biblical scholars call P (Genesis 1:1–2:4) and J (Genesis 2:5–3.2:4). After the formulaic and triumphal creation of everything in six days, followed by a day of rest, we read the version that places Adam and Eve in the Garden of Eden. Out of thousands of creation myths imagined by peoples everywhere, this double "just so" story produced long ago by an obscure Semitic people has won out over all others in the three principal monotheistic religions on Earth: Judaism, Christianity, and Islam. We come back to it again and again, less because it is ours than because it affords endlessly renewed meanings.*

The composite myth of Genesis 1–3 is so ancient and so dominant that many Bible readers do not notice that the creation story, once related, virtually disappears from the remaining one thousand pages of Scripture in the Old Testament.† It was St. Paul who, in a series of epistles, especially Chapter 5 of Romans, recast human history and theology by linking Jesus Christ, over the heads of all other prophets, leaders, and lawgivers, to Adam. The original man's transgression is now redeemed by the obedience of another man, God's incarnate Son. Jesus becomes the second Adam in a symmetrical pattern known in biblical study as typology. Christian faith proposes, among other things, an all-encompassing narrative unity.

Despite its familiarity, the creation story from Genesis is as invisible to many of us as air, or as our own personality. It surrounds us too closely. We cannot stand back in order to see it better. The Bible nowhere uses the word *fall* to designate what happened to Adam and Eve. The opening, or P, version over a period of seven days has the repetitive, invocational form of a hymn or poem. In the second, or J, version, we have suddenly moved in very close to a domestic scene where generic terms turn into proper names. God becomes the Lord God or Jehovah. Man becomes Adam—the Hebrew noun meaning "man," now particularized with a capital letter. Woman, meaning "taken out of man," finally becomes Eve, life-

---

*The extensive erudition and psychological keenness of two commentaries have guided me throughout this chapter: Arnold Williams, *The Common Expositor* (1948; especially Chapter VI, "The Fall"); and Howard Schultz, *Milton and Forbidden Knowledge* (1955).

†Job mentions Adam once (31:33) in a fleeting comparison. Jesus alludes to the *Genesis* story when he affirms monogamy and rejects divorce (Mark 10:6 and Matthew 19:4).

giver (Genesis 3:20). These leaps and turns survive translation nearly undiminished. The cast keeps growing. After forming Adam, the Lord God plants two particular named trees and imposes an interdict on the Tree of the Knowledge of Good and Evil. No such restriction applies to the Tree of Life, presumably because Adam is created immortal and does not need it—not yet. Its role will come later. Since Adam finds no helpmeet among the beasts and feels alone in Eden, God forms another and different living creature from his rib, "bone of my bones, flesh of my flesh." Adam names her woman. Now we have a God, two human beings, and two special trees as intermediate beings and props.

The third chapter opens with a jump cut to the serpent whispering in the woman's ear. The serpent is simply there, the tempter already in place, an unexplained occupant of the Garden—and of the human mind. The serpent appears to be the concentrated and symbolic remnant of an earlier religious age, before the Jews passed through the tumultuous shift from polytheism to monotheism. Nothing yet links the serpent to Satan or to the Devil. It is calmly insubordinate and categorically denies God's verdict of death for eating the forbidden tree. "Thou shalt not surely die" (3:4). The serpent tells the woman that, rather, the act will open their eyes and make them as gods. The woman eats and gives of the fruit to her husband. Everything goes by halves now. Adam and Eve start out innocent and immortal. The serpent claims that by eating the forbidden fruit, they will achieve divinity without losing immortality. He is half-right—that is, they attain insight into good and evil and at the same time they lose immortality. "And the Lord God said, Behold the man is become as one of us, to know good and evil: and now lest he put forth his hand, and take also of the tree of life, and eat, and live for ever . . . the Lord God sent him forth from the garden of Eden" (Genesis 3:22). Because they have become mortal, Adam and Eve must now be kept away from the Tree of Life. Prohibition did not work for the first tree. Banishment is the logical answer.

The cartoon figures and jagged episodes of Genesis provide an account of the first symbolic human encounter with taboo, both within us and outside us. The account conveys the powerful sentiment of "holiness and pollution . . . not yet differentiated," as Frazer describes it, a divided response of fascination and fear that characterizes the darkest stories of human life. The extreme economy of the Genesis Adam and Eve story (forty verses, about eight hundred words in English) has never been surpassed. Even without its later Christian link to the historical figure of Jesus of Nazareth, the story would probably have remained the creation myth for the three revealed monotheistic religions. It opens Hebrew Scripture from its earliest canon. It embeds its dramatic action in the universally desirable circumstances of a fruit-bearing tree in a lush garden of pleasures. With the clumsy directness of child actors, the cast enacts interlocking motifs of obedience and freedom, temptation and gullibility, sexuality and worship. Above all, the actions of both Adam and Eve show evil coming into the world through an inextricable combination of a preexisting outside force (the serpent) and of free choice in disobeying God's prohibition (seen clearly by Augustine). No other extant creation myth displays greater vividness and concentration in dealing with forbidden knowledge.

In comparison, the Prometheus story comes in many versions, dispersed across a series of episodes. In *The Greeks and the Irrational*, the classical scholar E. R. Dodds makes a strong claim. "Morally, reincarnation offered a more satisfactory solution to the Late Archaic problem of divine justice than did inherited guilt or postmortem punishment in another world" (Chapter V). Still, no religion or culture holding the doctrine of reincarnation has produced an establishing myth with the staying power of Adam and Eve. That fertile soil for interpretation and the taboo effect it creates help explain why it has given rise to more commentary, elaboration, and controversy than any other short passage of writing in all history.

We may find it surprising, therefore, that in the last half of the twentieth century one of the most eloquent and learned of Christian thinkers has responded with impatience to the Adam and Eve story. Unlike those of us who may have sold our imaginations to the big bang theory of the origin of everything or to the infinitely drawn-out minimalist drama of Darwinian evolution, Paul Ricoeur continues to honor Scripture. But time after time in his major work, *The Symbolism of Evil* (1967), he reveals his irritation with St. Paul's revival of Adam as the complementary figure to Christ and issues a testy challenge to Christian doctrine.

. . . it is false that the "Adamic" myth is the keystone of the Judeo-Christian edifice; it is only a flying buttress, articulated upon the ogival crossing of the Jewish penitential spirit. With even more reason, original sin, being a rationalization of the second degree, is only a false column. The harm that has been done to souls, during the centuries of Christianity, first by the literal interpretation of the story of Adam, and then by the confusion of this myth, treated as history, with later speculations, principally Augustinian, about original sin, will never be adequately told. In asking the faithful to confess belief in this mythicospeculative mass and to accept it as a self-sufficient explanation, the theologians have unduly required a sacrificium intellectus *where what was needed was to awaken believers to a symbolic super-intelligence of their actual condition.**

(*THE SYMBOLISM OF EVIL*, 239)

Apart from the incomprehensibility of the last clause, Ricoeur has stated a strange position. In an important essay a few years later, he comes back to the Adam and Eve story and discharges his im-

*I suspect that the middle sentence in this paragraph provided Elaine Pagels with the subject of *Adam, Eve, and the Serpent* (1988). Without a single reference to Ricoeur's powerful writing, Pagels covers much disputed ground in a short compass and stoutly defends the Gnostic position of untrammeled free will against any taint of Augustinian original sin. She evidently wishes that Adam and Eve would simply go away. "Perhaps the power of this archaic story, from which Christians have inferred a moral system, lies in its blatant contradiction of everyday experience" (128). Pagels cannot comprehend that, in addition to maintaining individual free choice, we need to attend to what everyday experience as well as the enduring myths imply about a positive force of evil in history and in ourselves, a force ready to tempt, to corrupt, to infect.

The Book of J (1990) by David Rosenberg and Harold Bloom solves none of these problems by defending the hypothesis that J, the Jahwist author of these sections of Genesis, was a woman at the court of King Solomon's son and successor. Bloom turns out to be another commentator impatient with the Adam and Eve story as written and seeking to demystify and to defuse it. "We have no reason to believe the serpent malevolent" (182), he writes, and goes on to state that he finds no candidates in Eden for culpability, except perhaps Yahweh himself, whose prohibition and temptation for his children was "a blunder" (183). A few pages later, Bloom sets out to deprive these events of their principal significance. "J's story of Eden . . . is anything but normative, as I have demonstrated. It is not a moral or a theological narrative, and asserts no historical status" (187). Like Ricoeur and Pagels, Bloom pays no attention here to *Paradise Lost.* Yet the reenactment of Adam and Eve in Milton's epic sweeps like a tidal wave over their attempts to dismiss the story.

patience on the "doctrinal stiffness" and "false logic" of original sin as Augustine defined it: both a juridical and a biological form of inherited guilt (*Conflict of Interpretations*, 1974). But this time, Ricoeur goes to great lengths to bring out the aptness and vividness with which the Genesis story dramatizes the double presence of election and seduction. "Evil is a kind of involuntariness at the very heart of the voluntary. . . .We inaugurate evil. It is through us that evil comes into the world. But we inaugurate evil only on the basis of an evil already there, of which our birth is the impenetrable symbol" (286). Ricoeur displays no naïveté about the supremacy of free will and does not doubt the real presence of evil as a force we are justified in calling Satan or the Devil.

Because he both responds to the drama of the Adam and Eve story and resists its doctrine, Ricoeur conveys a strong sense of the timelessness of the biblical verses. But his writings on the subject remain incomplete. For he does not take account of the one modern retelling of the story that is too important to be ignored, a version approaching a new Scripture. Milton gave to the Genesis narrative the epic dimensions and imaginative power of Homer and Virgil. After adequate attention to *Paradise Lost,* Ricoeur could not, I believe, have dismissed Adam and Eve as a "flying buttress" to the Judeo-Christian edifice. The opening chapters of Genesis have the simplicity of good wall paintings or tapestries. In contrast, *Paradise Lost,* behind its grand style, offers a scenario that an ambitious Hollywood producer would recognize without fail as the basis of a space-odyssey movie of the largest dimensions, one employing dazzling up-to-date special effects. One day, we may see that movie. Meanwhile, we have the poem that transforms the rudimentary Hebrew myth into a magnificent Christian epic.

In examining how Milton enlarges and enlivens the theme of forbidden knowledge from Genesis into a modern saga of self-discovery, I also wish to demonstrate that the poem carries remarkable appeal in its details, like the animated secular carvings that decorate Gothic cathedrals. Furthermore, Milton lived through a long political and moral conflict with his era and can communicate the excitement to us if we listen.

2. Milton's Version

Approaching fifty, Milton had lived through one of the most momentous decades in English history as an active pamphleteer and prominent public figure. In his writings, he advocated freedom of the press, freedom of religion, the right to divorce for incompatibility, and, most inflammatory of all, the right of subjects to put to death an unworthy king. In 1649, when the high court constituted by the Rump Parliament had the head of Charles I lopped into a basket in the name of the English people, Milton was appointed secretary for foreign tongues to the ruling Council and given an official residence in Whitehall. After the 1653 coup d'état, Cromwell appointed him as spokesman, a position similar to that of press secretary. The Puritan Revolution gained much of its intellectual vigor and style from Milton's classically trained mind.

By 1658, however, blindness, disillusionment with Cromwell, the death of his second wife, and a renewed poetic calling drew Milton back to private life. After the Restoration in 1660, his books were publicly burned and his life was in danger until, through the intervention of friends, he was included in the general amnesty. Milton had lived very close to the fire he had himself helped to light. Now, over fifty, he wanted to return to poetry and to earn a less scandalous reputation.

Since his early twenties, Milton had been seeking a suitably grand subject for a masterwork that would earn him a lasting reputation. Much of the time, he hesitated between the materials of classical epics and the more recent chivalric stories surrounding King Arthur. But there is good reason to believe that Milton felt deeply the pressure that Bacon, Descartes, and the new science were applying to modify the tradition of forbidden knowledge handed down from both antiquity and Christianity. And how could he respond to his recent revolutionary experience and to his earlier travels? During his continental tour in 1638, Milton had visited the aging and blind Galileo, who was living in enforced seclusion near Florence. Could this man of great learning be muzzled by a Pope? Milton found the indirect vehicle for these contemporary events in the oldest of all Old Testament stories.

There was not much precedent for new literary versions of the Adam and Eve myth. Scholars of scripture had produced a library of commentaries. At the end of the sixteenth century, Du Bartas had published a popular retelling in French verse that ran to fifty pages. It even had some success in English translation. But we should not underestimate the ambition and originality of Milton's project in two respects. He elevated Adam and Eve to the full dimensions of an epic subject, on a level with Homer and Virgil. And he imagined a modified and essentially modern version, which favors knowledge over the forbidding of it. Then he worked at it, blind and buffeted, for ten years.

The question of form also vexed him. An early manuscript sketches out a drama in five acts called *Adam Unparadised*. It shows the action of *Paradise Lost* already partly conceived in allegorical form. But he later chose to write an epic narrative poem conceived on a monumental scale. He multiplied the forty Old Testament verses (translated into their "authorized" King James version only in 1611) by a factor of four hundred to produce sixteen thousand unrhymed decasyllabic lines of diversified poetry. The epic narrative incorporates powerful dramatic scenes, a protestant and sometimes heretical theology of good and evil, a complex psychology that fluctuates from intensely human to unexpectedly playful, and a poetic diction like a powerful inboard motor that drives the story through wondrous cosmological and mythological spaces. *Paradise Lost* displays a cosmic imagination that produces episodes as grandiose as the scenes in the Sistine Chapel. The work also offers plain-spoken probings of domestic life comparable to those of Ingmar Bergman in films like *Scenes from a Marriage*. Above all, *Paradise Lost* merits a reading capable of releasing its "great unflagging voice," its "cantabile," as C. S. Lewis says of it. Milton, blind as a bard, dictated every one of these lines in his resolve to compose an epic not just for one nation, like Homer's and Virgil's, but for all humanity.

In order to make sure that his readers follow the story, Milton supplies a page-long "argument," or summary, for each of his twelve books. Following the same impulse, I now propose a single synopsis of the poem, highlighting the central events concerning Adam and Eve. This way, I believe, the reader can grasp the movement of the narrative behind the frequent flashbacks, anticipations, digressions, and authorial interventions. This truncated version also smuggles in a certain amount of commentary and interpretation.

THE PLOT

After a high-decibel invocation addressed both to the pagan Muse and to the Christian Spirit to help him "soar" higher than any previous poet, Milton opens not with the creation story but with the kind of world-shaking events he had himself just lived through and survived: a rebellion and its collapse. Cast out of heaven by the Lord as punishment for his attempted revolt, Lucifer-Satan the Arch-Fiend and his followers regroup in Hell and plot revenge on a new world rumored to have been created by the Lord elsewhere in the universe (I). Satan ventures forth alone on a great intergalactic voyage to discover the whereabouts of Paradise. Approaching the Gates of Hell in order to leave it, he finds them guarded by two unspeakable monsters. One is Sin, a sorceress sprung fully formed out of Satan's head at the instant he first conceived envy for the Son of God—a pastiche of Minerva born of Zeus' head. The other is Death, the odious offspring of Satan's incest with Sin.* With her "fatal key, / Sad instrument of all our woe," Sin unlocks the Gates of Hell and liberates Satan to pursue his mission (II).

Looking down from on high, God sees Satan approaching Adam and Eve in the Garden of Eden and foresees that Man, created strong enough to resist temptation yet free to transgress, will fall. By an unelucidated paradox this prophecy does not constitute predestination, does not determine events. The Lord explains to his Son that, unlike the fallen angel Satan, "self-tempted, self-depraved" (III 130), Man, deceived by Satan, will find grace. The Son offers himself as the instrument of this glorious act (III). Meanwhile, Satan has a brief twinge of doubt and remorse over his prideful rebellion against God, then a second recoil when he sees Adam and Eve. He could almost love and pity the gentle, comely pair and he overhears that they live under "one easy prohibition" (IV 433), not to eat the fruit of the Tree of Knowledge. But the spectacle of their innocent connubial love, "Imparadised in one another's arms" (IV 506), fills him with torments of envy. He resolves to destroy their happiness. The Lord's agents catch Satan in the

---

*By inventing these episodes, Milton separates the origins of Sin and Death from any act of Eve or Adam, who, rather, draw down on their heads the fate of these preexisting figures. Thus Milton modifies the Augustinian doctrine of original sin.

shape of a toad whispering into Eve's ear while she sleeps. He has to retreat temporarily (IV).

In the morning, Eve recounts her "uncouth dream" of being tempted by an angel to eat of the forbidden tree. Puzzled by this unexplained manifestation of evil, Adam reassures her, and they pray together. Then they welcome as an unexpected guest in Paradise the Archangel Raphael. He is sent by God to forewarn them of their free condition, permitting both obedience and disobedience, as in the case of Satan. Adam inquires of that story, and Raphael tells at great length the events of Lucifer-Satan's rebellion and his defeat by the Son after a great battle (V–VI). At Adam's request, Raphael goes on to describe the creation of the World in six days and a sabbath (VII). When Adam asks about cosmology and celestial motion—that is, the Copernican debate—the angel draws the line at talking about these "things too high" (121). Accepting this admonition, Adam relates in a long flashback his life since his own creation, his conversations with God, the creation of Woman for companionship, and the transport of passion he experiences in the presence of her beauty and in their guiltless nuptials. Raphael warns Adam against subjugation to passion and reveals that Adam is free to stand or fall in the face of temptation (VIII).

In an eloquent second invocation, Milton regirds himself for the central events of his story and affirms them as more heroic than either Greek and Roman epics or the modern chivalric tales of gorgeous knights in battle. At her own suggestion, Eve is working apart from Adam in the Garden; she comes upon Satan, now in the shape of a serpent. He claims that eating the fruit of the Forbidden Tree has given him the power of speech. His subtly reasoned temptation speech suggests that knowing evil will help her to shun it. A just God, he argues, could never punish by death. She eats and feels unparalleled delight. But still fearing she will die, and therefore jealous of Adam's future without her, she offers the fruit to him. Eve draws Adam into her own death. Out of love for her, knowing the consequences better than she does, Adam also eats. Straightway, their innocent love changes into the dalliance of guilty lust. They feel shame and fall into mutual recriminations (IX).

God the Son descends to Earth to pass judgment on Adam and Eve. Sin and Death (forming with Satan a competing trinity) enter the world now that Satan has prevailed. Adam first protests the injustice of his fate thrusting him into an existence he never asked for; then he accepts his responsibility and wishes for immediate death. His new-formed conscience laments the consequences of his act for all human posterity. Adam firmly resists Eve's proposal of suicide, and they are reconciled to each another, to their long day's dying together, and to hope through acceptance and prayer (X).

The Archangel Michael arrives to announce that they must leave Paradise. On a hilltop, he gives Adam an illustrated preview with commentary of the future course of mankind up to the Flood (XI). The visions culminate in the incarnation of the Son and the redemption of man from sin and death, followed by the parlous events of modern times. Adam marvels that all this good should one day issue from his original evil act. Michael leads the pair out of Paradise with both sorrow and hope in their minds (XII).

FOR FOUR HUNDRED PAGES, Milton's ten-syllable lines fill the columns before our eyes in a uniform visual pattern that reveals nothing about tone, pace, and portent. It takes an articulated, spoken reading to shape *Paradise Lost* into the repertory of moods and styles Milton deployed during the ten years of its composition. He could shift from stentorian-prophetic to the downright folksy. In Book V, Eve spreads out a generous *déjeuner sur l'herbe* for their heavenly visitor, the Archangel Raphael. They get to talking, though, and the narrator slips in a sly post-Edenic joke. "A while discourse they hold; / No fear lest dinner cool" (V, 395–96). Later Milton in his own voice will tell us that the English "cold / Climate" made composition of the poem very difficult (IX, 44–45). We imagine his chilblains. Raphael, encouraging Adam to tell his creation story, implies that as a busy archangel he was out of town on a business trip that day.

> *"Say therefore on;*
> *For I that day was absent, as befell,*
> *Bound on a voyage uncouth and obscure,*
> *Far on excursion toward the gates of Hell . . ."*
>
> (VIII, 228–31)

God himself is depicted as laughing at men's "quaint opinions" (VIII, 78) about the layout of the heavens. A few pages later when Adam complains to the Lord of the lack of human companionship in Paradise, God surely grins at him.

> *"What thinkst thou then of me, and this my state?*
> *Seem I to thee sufficiently possessed*
> *Of happiness, or not? who am alone*
> *From all eternity . . ."*
>
> (VIII, 403–6)

In general, one must recognize in Milton a special pre-Joycean language that exhibits its Latinate origins in a liberated word order and revels in reversion to the root meanings of words. Enjambment, elision, and repetition constantly vary the flow of his blank verse. This master poet in Latin and Italian as well as English, who had been rhyming skillfully for thirty years, now barred rhyme from his most ambitious work. His opening note on the verse summarily dismisses rhyme as "the invention of a barbarous age"; Milton permits himself only seventeen couplets or near couplets—approximately one per one thousand lines. Two of them have an important function: They flag the central acts of the book and dramatize the cosmic reactions first to Eve and then to Adam as each eats the forbidden fruit.

> *Forth reaching to the fruit, she plucked, she eat.**
> *Earth felt the wound, and Nature from her seat*
> *Sighing through all her works gave signs of woe*
> *That all was lost.*
>
> (IX, 781–83)

*Many modern versions change the word "eat" to "ate." Seventeenth-century pronunciation of these words is uncertain and may have resembled *et.*

Earth trembled from her entrails, as again
In pangs, and Nature gave a second groan.

(IX, 1000–1001)

Here then is Milton's wager: He will hang everything—the whole human condition and his own reputation—on the Adam and Eve story. Out of an original Hebrew version as primitive as a few stick marks on the wall of a cave, he unfolds a drama of epic proportions and embeds it in an account of all previous and subsequent history, including the religious and political struggles of his time (XII, 507–37). Why is Milton so confident that this lowly tale will outshine and outlast the magnificent deeds of ancient heroes and of knights errant in combat?

I believe it is because within the expanded action Milton can focus on the question of *knowledge*—knowledge proffered and knowledge forbidden. The following pages will document that claim. I can also illustrate it with one touching passage. Not a third of the way through the poem, the poet looks down on the couple after they have experienced the full physical delights of innocent copulation. He blesses their happiness. And for a moment, the poet appears to want to hold back their inevitable fate of knowing more, knowledge that will end their blessedness and complicate everything:

Sleep on
Blest pair; and O yet happiest if ye seek
No happier state, and know to know no more.

(IV, 774–76)

Milton's sounds reinforce the scene and the theme. The blissful *O* rhymes with a hovering, ominous off-stage *woe. Know* dances a slow, suggestive saraband with *no*. The whole action balances on knowing and not knowing. The lines beg to be said aloud, to be sung.

3. "OF KNOWLEDGE WITHIN BOUNDS . . ."

Why do Adam and Eve fall from their paradise of innocence and immortality? What more or other could they possibly want?

> *Of Man's first disobedience, and the fruit*
> *Of that forbidden tree, whose mortal taste*
> *Brought death into the world, and all our woe . . .*
>
> (I, 1–3)

The opening lines establish a priority of themes that has had lasting authority among readers of *Paradise Lost.* C. S. Lewis declares uncompromisingly that the Fall represents an act of disobedience; the apple has no intrinsic importance even though Eve and Satan may believe so. In other words, in the term *forbidden knowledge* closely associated here, the emphasis falls on the word *forbidden.* Eve and Adam act in large part out of perverseness, an unwillingness to obey the contract by which they have been granted residence in the Garden. They are just too ornery or too curious or too spoiled to tolerate any prohibition at all.

I believe this interpretation is too restrictive. In order to do justice to Milton's version, we must examine some passages that precede Book IX, where the actual Fall takes place.* During the four books that narrate his gossipy fraternizing with Adam and Eve, the Archangel Raphael has a friendly mission to perform for the Lord concerning Adam: to warn him to "beware / He swerve not" (V, 236–37). But before the angel can carry out his mission, Adam takes the initiative. He starts asking questions. It is as if human waywardness here unexpectedly springs full grown from Adam's head as Sin sprang from Satan's. Satan was driven by envy of God's Son; nothing seems to cloud Adam's contentment except that his speech is "wary" (V, 459).

*We should also remember that disobedience to the king and to his divine authority was the offense for which the Restoration condemned the Puritan Revolution and Milton's participation in it.

Thus when with meats and drinks they had sufficed,
Not burdened nature, sudden mind arose
In Adam, not to let the occasion pass
Given him by this great conference to know
Of things above this world, and of their being
Who dwell in Heaven . . .

(V, 452–55)

Already looking beyond the Paradise conferred on him, Adam frames a question about how this life compares with life in the entourage of the Lord in Heaven. Raphael says he'll learn the answer "if you be found obedient" (V, 501). What can that possibly mean, since we are so blissful? asks Adam. The angel explains patiently that Adam is free to lose by disobedience the happy state given by the Almighty. Angels share the same condition. "Freely we serve, / Because we freely love" (V, 538–39). Remembering Satan's recent insinuation into Eve's fancy in a dream, Raphael adds that some have indeed fallen by disobedience from a "high state of bliss into what woe!" (V, 543). Knowing nothing of all this, our ancestor, feeling "some doubt within me move" (554), asks for "the full relation" (556). There's plenty of time he adds helpfully. Then in a revealing fourteen-line preamble to the story of Satan's rebellion and the War in Heaven, Raphael not only reflects that it will be difficult to narrate these angelic events; he also wonders, as he begins to unfold them, if they are "perhaps / Not lawful to reveal?" (569–70). What kind of a moral lesson will Satan's story provide for Adam? Raphael has no instructions on this point. A responsive reader will introduce here a lengthy pause of indecision on Raphael's part.

Raphael finally answers his own question about forbidden knowledge according to a principle of freedom proclaimed (with careful reservations) in Milton's pamphlet *Areopagitica* on censorship and the press. Ignorance of evil implies lack of free choice, a "blank virtue" and a "puppet Adam." These libertarian arguments from *Areopagitica*, though unexpressed at this point in *Paradise Lost*, loom large between the lines and explain the angel's willingness to tell the full story of Satan's disobedience. Raphael dismisses his own hesitations about educating Adam by saying that it will all be "for thy good" (570). The cautionary tale of Satan's rebellion and defeat consumes thirty pages, a whole new book, and a cast of thousands.

Book VII opens with a convenient summing up after the long insert story. We are asked to believe the narrated events both had their intended warning effect on the happy pair and had no effect. For Adam comes back for more. It is a key passage, subtly written. (In the first line, Milton invokes his muse.)

> *Say Goddess, what ensued when Raphael,* 40
> *The affable Archangel, had forewarned*
> *Adam by dire example to beware*
> *Apostasy, by what befell in Heaven*
> *To those apostates, lest the like befall*
> *In Paradise to Adam or his race,*
> *Charged not to touch the interdicted tree,*
> *If they transgress, and slight that sole command,*
> *So easily obeyed amid the choice*
> *Of all tastes else to please their appetite,*
> *Though wandering. He with his consorted Eve* 50
> *The story heard attentive, and was filled*
> *With admiration and deep muse, to hear*
> *Of things so high and strange . . .*
> . . . .
> *Whence Adam soon repealed*
> *The doubts that in his heart arose; and now* 60
> *Led on, yet sinless, with desire to know*
> *What nearer might concern him, how this World*
> *Of Heaven and Earth conspicuous first began,*
> *When, and whereof created, for what cause . . .*
> . . . .
> *Proceeded thus to ask his heavenly guest.*
>
> (VII, 40–69)

Don't do what Satan did, says the parable, provoking "deep muse" (52) in Adam. We also learn that his appetite is already "wandering" (50), an adjective underlined by its placement in line and sentence. Soon, being sinless, he repeals these "doubts" (60)—hesitations, reflections, questionings. But even in his innocence, he desires to learn more and in the following question—his fourth during this picnic—includes an elaborate explanation for pestering the archangel further. Adam pushes Raphael to tell the whole creation story:

> *. . . if unforbid thou may'st unfold*
> *What we, not to explore the secrets ask*
> *Of his eternal empire, but the more*
> *To magnify his works, the more we know.*
>
> (VII, 94–97)

Don't think I'm prying; I seek only better ways to glorify God. Adam's argument for more revealing stories resembles Bacon's in favor of scientific research in *The Advancement of Learning.* Patiently, Raphael goes along, with a gentle demurrer that he has orders:

> *. . . to answer thy desire*
> *Of knowledge within bounds; beyond abstain*
> *To ask, nor let thine own inventions hope*
> *Things not revealed . . .*
>
> (VII, 119–22)

Milton-Raphael is willing to tell all the good tales; headquarters does not want anyone to forget the "bounds" of knowledge. The creation story that follows, beautifully illustrated a century later by William Blake, occupies only twelve pages. Adam, almost hypnotized by Raphael's voice, has now presumably heard everything he wants to know. But no. Book VIII begins with an almost-comic sequence. Incorrigible, Adam says that "something yet of doubt remains" (VIII, 13) about the celestial motions of such a disproportionate number of stars just to light tiny Earth. This time, he provokes two strong reactions. Eve gives up and walks off into the Garden to escape the extended disquisition she expects (40ff.). Raphael, firmly now, lowers the boom on further discussion, thus allowing Milton to withdraw from taking a position in the still-raging Copernican controversy. The lengthy wrist-slapping, so long in coming, tells Adam not to presume to know what lies beyond his reach (89ff.) and closes with famous lines counseling sobriety and humility.

> *Heaven is for thee too high*
> *To know what passes there; be lowly wise:*
> *Think only what concerns thee and thy being;*
> *Dream not of other worlds, what creatures there*
> *Live, in what state, condition, or degree,*
> *Contented that thus far hath been revealed*
> *Not of Earth only but of highest Heaven.*
>
> (VIII, 172–78)

Milton's eager narrator tells us immediately that Adam is "cleared of doubt" (179), satisfied. But the unfallen Adam is not so compliant a believer as that. Adam says he will live "the easiest way," free of "perplexing thoughts" (183) . . . *unless* . . .

> *. . . unless we ourselves*
> *Seek them* [cares], *with wandering thoughts and notions vain.*
> *But apt the mind or fancy is to rove*
> *Unchecked, and of her roving is no end;*
> *Till warned, or by experience taught, she learn*
> *That not to know at large of things remote*
> *From use, obscure and subtle, but to know*
> *That which before us lies in daily life,*
> *Is the prime wisdom; what is more, is fume,*
> *Or emptiness, or fond impertinence. . . .*
>
> (VIII, 186–95)

"Wandering," we already know, means trouble. Here is a prelapsarian Adam-Tartuffe slyly inverting the situation to suit his purposes. He finds the word *experience* (190) to turn the trick. Obedience and humility are fine, he tells Raphael, except for the fact that the imagination tends to rove out of control. Stern warnings, like the one just given, may help. But worldly experience will teach us better and faster to be "lowly wise" and to avoid "notions vain." "Experience" emits a whiff of rebellion against constituted authority. Adam, still presumably innocent, is preaching a very Blakean sermon on how the road to innocence passes through experience. And with that theological dilemma, Adam, not Raphael, ends the discussion of the perils of knowledge and offers to tell his own story.*

As later passages will make clear (XI, 807, 988), behind experience looms disobedience, the Fall, and all our woe. In a strongly argued interpretation of this passage and others, Millicent Bell tracks how "an instant of waywardness" in Adam develops into the "lust of forbidden knowledge." Exactly. And then in the next and central book, Book IX, having focused on Adam through four books of conversation with Raphael, Milton will follow Genesis and have Eve, rather than Adam, act out the subversive thoughts expressed primarily by him. Furthermore, Eve's dream or fancy at the opening of Book V, a nonbiblical and nontraditional foreshadowing of the temptation scene, gives her a role in the lengthy preparations. Milton keeps the woman's role central to the action of forbidden knowledge.

Now at the climax of the drama, Satan as serpent dismisses God's covenant in four words ("ye shall not die," IX, 685) and goes on to seduce Eve by recapitulating arguments already planted by Adam. God will surely approve of her courage in scorning death in order to achieve a happier life. (All of *Faust* and the character's striving lie here in germ.) And how could God oppose knowledge gained from the Tree of the Knowledge of Good and Evil? Certainly one should know the good. Then the clincher: ". . . of evil, if what is evil / Be real, why not know, since easier shunned?" (IX, 698–99). Satan is repackaging Adam's argument: Experience protects us from evil better than mere warnings against it. After greedily ingorging the fruit, Eve gives fervent and idolatrous thanks, first to the Tree of Knowledge and next to "Experience . . . / Best guide" (807–8).

With her new knowledge, Eve now fancies herself superior to Adam and freer than he. In the changed situation, she is tempted to lord it over him. But second thoughts strike her immediately and without mercy. For Adam has already reminded her in his speech

*In these moments of restlessness, Adam's tone and vocabulary resemble Ulysses' (e.g., *esperienza*) in the episodes Dante adds to his story (see p. 25).

that "God hath pronounced it death to taste that tree" (IV, 427). Her imagination leaps swiftly ahead of her euphoria.

> *But what if God have seen*
> *And death ensue? then I should be no more,*
> *And Adam wedded to another Eve*
> *Shall live with her enjoying, I extinct;*
> *A death to think . . .*
>
> (IX, 826–30)

Unable to tolerate the prospect of immortal Adam surviving her death and enjoying another woman, Eve resolves to induce him to eat the same fruit and share death with her. These are not beautiful thoughts. C. S. Lewis goes so far as to say Eve murders Adam. His infatuation with her makes him easy to persuade, even though he is "not deceived" (998) by her final hollow argument.

> *On my experience, Adam, freely taste,*
> *And fear of death deliver to the winds.*
>
> (IX, 989–90)

She has slyly picked up his word—*experience.* Throughout this scene, Eve has played the part of Adam's surrogate, deputized with the full power of the one trait that propels the whole drama, the trait with which Adam retains Raphael through several renewed conversations, and for which our vocabulary provides the singularly mild word: *curiosity.** The principal yeast with which Milton leavens the forty stark verses of Genesis 3 into the great loaf of his epic poem is *libido sciendi*, "the lust to know." It works constantly in the words and thoughts of Adam and, at the moment of her most dire crisis, it guides the actions and responses of Eve. Millicent Bell was right to plot curiosity all the way from "an instinct of

*The entries in *The Complete Oxford English Dictionary* for *curiosity* and *curious* trace a sequence of overlapping meanings: originally attention to detail, carefulness; then, up to the seventeenth century, blamable inquisitiveness, "adultery of the soul," "spiritual drunkenness"; and finally, the neutral or positive modern sense of eagerness to know and to learn.

waywardness"—a child's idle toying with the world within reach—to "the lust of forbidden knowledge"—a drive carrying a strong element of perverseness and a penchant for transgression.*

In the twentieth century, which honors the bold forays of science into the mysteries of nature and the alluring possibility of space exploration, curiosity strikes us far more as the beginning of wisdom than as the beginning of sin. In seventeenth-century England,

* The full significance of Milton's interpretation of the Fall is brought out by comparison with that of a great poet closer to the Middle Ages. In the *Paradiso*, Dante asks Adam what was the nature of the first sin to provoke God's wrath. Adam's three-line answer sets aside gluttony or curiosity (satisfied by pleasure or knowledge gained by eating the forbidden fruit) in favor of sheer disobedience of God's prohibition.

> *Know now, my son, the tasting of the tree*
> *was not itself the cause of such long exile,*
> *but only the transgression of God's bounds.*
>
> (PARADISO, XXVI, 115–17, TR. MARK MUSA)

Dante's "transgression" is sternly categorical compared to Milton's "wandering" and "experience." Those words allow for the human content of sin: knowledge to satisfy a lust of the mind, curiosity.

There will always be more. In a section entitled "Human Knowledge" in *Nosce Teipsum* (1599), Sir John Davies narrates how tasting the forbidden fruit in search of knowledge made Adam and Eve blind. I quote three stanzas because their earthy rhymed beauty contrasts vividly with the Latinate grandeur of Milton's lines and because Davies, too, finds the unexpected word *experience* (in the first stanza, which also rhymes "know" with "woe").

> *For then their minds did first in passion see*
> *Those wretched shapes of misery and woe,*
> *Of nakedness, of shame, of poverty,*
> *Which then their own experience made them know.*
>
> *But then grew reason dark, that she no more*
> *Could the fair forms of good and truth discern;*
> *Bats they became, that eagles were before,*
> *And this they got by their desire to learn.*
>
> *But we, their wretched offspring, what do we?*
> *Do not we still taste of the fruit forbid,*
> *Whiles with fond fruitless curiosity*
> *In books profane we seek for knowledge hid?*

Davies sustains a fine diction and draws his moral more directly than Dante and Milton—and very gracefully. I have modernized the language of the version given by Hershey Sneath.

the dawn of the modern scientific era, a swarm of disputations gave prominence to the *presumption* of human knowledge. Howard Schultz's book *Milton and Forbidden Knowledge* describes those disputes and informs us that Milton was familiar with Bernard of Clarivaux's motto—"Curiosity is the beginning of all sin"—and with the apostle Paul's warning—("*sapere ad sobrietatem*") ("learning guided by sobriety"). Tillyard points out that Milton's commonplace book carries three entries under the heading "Curiosity." Puritan preachers referred wrathfully to "Adam's disease." Montaigne's disciple Charron, in a book entitled *On Wisdom*, argued strongly for ignorance. Milton wrote very much in the midst of controversies between old sobriety and new science.

By the time we reach the Fall in Book IX of *Paradise Lost*, the categorical "disobedience" of the epic's opening line has been colored and attenuated by two traits depicted as winningly human: curiosity and the appeal to "experience." They explain the actions of both Adam and Eve without excusing them. There stands the alluring and mysterious Tree of Knowledge, flaunting at every moment its special status. Could the privileged residents of the Garden simply ignore it as they were instructed? There is one more factor in the story to explain why the restriction on eating the fruit could not remain, as Adam said in his first speech to Eve, "one easy prohibition" (IV, 433).

I shall call it "the Wife of Bath effect." This subtle yet powerful human trait underlies many of these discussions of forbidden knowledge and combines several unwelcome yet familiar elements of our condition. We are discontent with our lot, whatever it is, just because it is ours. We covet what is not ours because it represents otherness. Following Montaigne, I have called this combination of perverse impulses "soul error" and identified it as a vital motif in the works of Proust and many other writers. To this odd yet common dissatisfaction with ourselves even when we may be happy, a further complication can be added: a constraint or prohibition. It only makes things worse. The great narratives of all time explore this conflict as it inflames love, adventure, war, crime. The most succinct telling of the tale in all literature occurs in the seven-word line from Chaucer I have used as an epigraph for this book: "Forbede us thyng, and that desiren we."

Death-defying feats draw many contenders. The higher the wall, the greater the challenge. Some women are attracted to a criminal rapist, some men to a known "man-eater." Children must, like Bluebeard's wife, play with the one object they are told not to touch. The imp of the perverse that lurks in our restless minds may lead to self-injury and self-destruction. It is as if the concatenation of steps were as inexorable as the playful psychological challenge: Do *not* think of a pink elephant in a blue desert. The prohibition creates a vacuum into which our freedom of will seems to be sucked by a strong natural law. Only an equal counterattraction can save us from what Milton called "the instinct of waywardness."

Unlike the lines quoted about evil ideas leaving "no spot or blame" in Eve's mind, the Wife of Bath effect emphasizes the "forbid" side rather than the "know" side of forbidden knowledge and recognizes the perverse pull exerted on our frail moral faculties by any prohibition. Milton comes close to implying extenuating circumstances for "Man's first disobedience." Without embracing that heresy, he makes clear that Adam and Eve have learned their lesson.

4. THE DOWNWARD PATH TO WISDOM

At crucial junctures of *Paradise Lost*, Milton explores the questions of freedom and government that inspire his ringing declaration in *Areopagitica*. With all its dodges, that pamphlet goes further than any earlier document to defend freedom of speech and publication on the basis of individual free choice. The argument about the free circulation of ideas appears once fairly early in *Paradise Lost* and remains as a troubling motif through the subsequent exposition, crisis, and denouement. In Book V, after Eve describes her dream of being tempted by a gentle-voiced angel, Adam broods over the source of "This uncouth dream, of evil sprung" (V, 98). For Eve was "created pure" (100). Not finding a simple answer to this echo of the question that opens the epic (What caused the Fall? [I, 27ff.]), Adam starts a disquisition on "Fancy" and on faculties that compose the mind or soul, namely reason and feeling. The latter is prone to produce dreams, but Eve, Adam says, need not be disturbed by her strange dream.

> *"Evil into the mind of god or man*
> *May come and go, so unapproved, and leave*
> *No spot or blame behind; which gives me hope*
> *That what in sleep thou didst abhor to dream*
> *Waking thou never wilt consent to do."*
>
> (V, 117–21)

It is necessary but not easy to sort out the contradictions and paradoxes lodged in these lines, which are assigned the function of clearing the air before Adam and Eve's innocent prayers (V, 209). Because of the connections with Mark 7:15*, with Dante's dreams in *Purgatorio,* and with Milton's own *Areopagitica,* we understand that much is at stake here. From the above passage and others related to it, we can infer four forms or stages of knowledge.

Milton never lingers long over the first state of pure ignorance or innocence. Both Eve and Adam display traits of curiosity, vanity, deviousness, which hover tantalizingly between unselfconsciousness and corruption. The second form of knowledge comes through fancy or dream, a purely imaginary encounter with worldly actions, as in Eve's dream. These five lines assure us that such fanciful encounters with evil leave no spot; they imply not infection but something approaching a catharsis theory of imagination—a vicarious adventure followed by cleansing. Still, the passage gently resists the interpretation I have just given it. Does "mind" mean fancy? Or reason? Or both? Adam says "abhor"; Eve's account (V, 29–94) reveals that her first temptation in an interrupted dream inspired in her both "horror" and "exaltation." Is she still spotless?

The third step of knowledge is full experience, the actual doing that commits reason, fancy, and all the senses. Where fancy by itself, the entertainment of ideas or images, remains blameless, experience entails the consequences of free choice and responsibility. In Book

---

*After rebuking the Pharisees, Jesus says to the people: "There is nothing from without a man, that entering into him can defile him: but the things which come out of him, those are they that defile the man" (Mark 7:15). In context, he means that unclean foods pass through us without doing harm, but unclean words and deeds reveal the corruption within us.

IX, the full experience of eating the forbidden fruit brings about the Fall. Both Adam discussing his "doubts" with Raphael (VIII, 190) and Eve still all aglow from eating the fruit (IX, 807) explicitly name "experience" as the great teacher. What can it teach beyond itself? Beyond bliss and pain? In this case, beyond mortality?

For this fourth stage, Milton uses another traditional word, more classic than Christian, that now encompasses knowledge of good and evil. Raphael's advice to Adam during their long conversation before the Fall comes too soon: ". . . be lowly wise" (VIII, 173). For true wisdom arrives only at the end of the epic story, when experience has done its work, after Adam has conceded, "Henceforth I learn, that to obey is best" (XII, 561). Then the Archangel Michael pronounces what is essentially the verdict and blessing of this long trial.

> *This having learned, thou hast attained the sum*
> *Of wisdom; hope no higher . . .*
>
> (XII, 575–76)

These benign lines carry a conclusiveness absent from their earlier version, "know to know no more" (IV, 776), uttered in vain over the sleeping couple while they are still in the first state of complete innocence. A little earlier in the last book, after hearing Michael relate the incarnation and atonement story, Adam marvels ecstatically at God's goodness "That all this good of evil shall produce" (XII, 470). In the Christian story, the Fortunate Fall interprets Adam's sin as the action that permits redemption by the second Adam, Jesus Christ. In vivid filigree behind the theological meaning of Eden, Milton narrates a secular story about a legendary yet very human couple who move through four stages of knowledge: innocence, fancy or dream, experience, and wisdom. We can read *Paradise Lost* as a tale about the downward path to wisdom, a path that must lead through the experience of sin.

Let us pause a moment to reflect again on how *Paradise Lost* embraces, enlarges, and deepens the bare action of Genesis 3. In one illuminating respect, Milton's *Paradise Lost* stands in relation to Genesis as Aristotle stands to Plato. Plato banished the poets as agents of infection who excite our passions and our senses. Aristotle's *Poetics* finds a place for poets as agents of a catharsis that enlarges our moral understanding. In a loosely parallel fashion, Genesis banished Adam and Eve to eternal penance for their disobedience. *Paradise Lost* permits them to contemplate the eventual surpassing of their sin by true moral understanding and by Christian redemption. The Lord says that by eating of the Tree of the Knowledge of Good and Evil, "thou shalt surely die" (Genesis 2:17). The serpent says to Eve that by eating of the Tree, "your eyes shall be opened, and ye shall be as gods, knowing good and evil" (3:5). Milton's epic retelling shows how both can be right. Mortality and knowledge together form our lot. And in both stories, the prohibition is necessary; it thickens the plot, according to the Wife of Bath effect. Something must be there to set the limit—divine prohibition, or civil laws, or traditional morality, or the inner voice of conscience. The sway of one or more of these forces enables us to turn experience into wisdom. Without them, we sink into selfishness and self-indulgence.

The carefully controlled experience of evil in the Eden story lies close to the practice of vaccination. A restricted dose of disease or infection stimulates an immune reaction. The epigraph Baudelaire found in d'Aubigné for *The Flowers of Evil* transposes the medical principle of vaccination to the moral-intellectual realm: "For virtue is not the fruit of ignorance." The line also recapitulates the central argument of Milton's *Areopagitica*.

These home truths about innocence and experience, about fancy and wisdom, and about prohibition cannot be expressed in a few lines of commentary that try to extract the essence of a legendary story. There is no substitute for Genesis 3 in its stark suggestiveness, nor for *Paradise Lost* in its extended metamorphosis and dramatization of all that has grown out of the original. They vie with one another undiminished and make rival claims on our imagination in ways that illuminate both the riches of literary history and the long struggle to assemble a moral order.

It is time now to look at the moment near the end, when Adam interrupts Michael's foretelling of Abraham and Moses, the law and the covenant, to say, "Now first I find / Mine eyes true opening" (XII, 273–74). During the scene of the Fall, the serpent tells Eve her eyes will be opened (IX, 706–8). She says the same thing to Adam (865–66), and after he eats the fruit, the narrator repeats it (1053). But the following lines reveal that at this point their eyes are opened only "To guilty Shame" (1058). When Milton writes in the last book about Adam's eyes' "true opening," the context tells us that the true knowledge implied is scriptural, revealed. Adam goes on to say that this revelation of the future is a special favor for him "who sought / Forbidden knowledge by forbidden means" (XII, 279–80).

Milton is not standing Genesis on its head. For all the reverberations of rebellion and undertones of discord that his narrative sets off, he never ceases both to sympathize with and to excoriate the sin of pride in the form of *libido sciendi*. We want to know too much. We feel the pull of the Wife of Bath effect. This immense poetic and theological testament, devoted to restaging the greatest story ever told, incorporates warnings against proud knowledge as stringent as the Tower of Babel episode and Candide's "Let us cultivate our garden."

We should not be surprised that a great work of Christian faith produced in the turbulence of seventeenth-century England should carry in its recesses and its structure, along with the tireless conspirator Satan, elements of doubt directed primarily toward the abuse of human freedom and the faculty of fancy. Across the Channel, Descartes was using systematic unsparing doubt as a method to clear the ground for inductive thought, leaving in place only as much of God as was necessary to start the motor of being. Coming from the other direction, Milton wished to reestablish the great European religious tradition in sturdily Protestant terms. Yet the two human characters he created to enact that story display a faith in the Lord sensibly alloyed with doubt in the form of inextinguishable curiosity. The tale of Adam and Eve and the serpent offers us many latent messages about disobedience, sexual concupiscence, and male superiority.

But the center is not located there. Milton almost allows Satan to steal the starring role and the moral high ground. But Satan's resourceful and defiant performance remains a matter of choosing the right tactics to corrupt Adam and Eve in their enviable Paradise, not of finding the right conduct for human life. Writing at the historical moment when Descartes and Pascal represented the poles of philosophical thought in France, Milton gave his epic poem unparalleled scope by incorporating into it two corresponding sets of opposites: knowledge and ignorance, doubt and faith. Their pincers close on the central paradox of what we now have reason to call forbidden *experience*. Milton puts the word into Adam's and then Eve's mouth, much as Dante puts it in Ulysses' mouth, to designate action leading first possibly to sin and later to wisdom and salvation. We cannot abstain from living. We cannot eliminate the Wife of Bath effect. But Milton is equally clear about Archangel Raphael's injunction to Adam not to reject experience and knowledge, but to limit them: "Be lowly wise" (VIII, 173).

CHAPTER III

FAUST AND FRANKENSTEIN

■

1. THE FAUST MYTH

In the flamboyant figure of Satan, Milton alludes to the momentous events through which he himself had just lived: a revolution that failed. History would bring about several more. But despite readers' enthusiastic response to Satan's role, the central narrative of *Paradise Lost* rehabilitates one of the oldest stories from Hebrew mythology. Milton did not invent a new plot.

After cohabiting for many years with the corpus of Western literature, I sometimes wonder if it all could be reduced to a few simple stories. James G. Frazer and his epigone Joseph Campbell attempted such a synthesis for myths and legends. On his singing-reciting tours, the poet Carl Sandburg used to utter with banjo accompaniment what he called the shortest poem ever written: "Born. Troubled. Died." Others have proposed thirty-six dramatic situations. The folklorist Vladimir Propp thought he was accomplishing something worthwhile by identifying in Russian folktales 31 functions and 151 elements, with a mathematical symbol assigned to each. The slow collective crystallization of popular stories into a handful of myths reveals some of the shapes our lives may take and the yearnings they express or repress. One of the distinguishing features of our Western collection of myths is that most of them come from ancient sources—Egyptian, Greek, Judaic, Near Eastern. The number of postclassical myths is so limited that I can identify only two that have emerged in the last thousand years.

The first consists of the extravagant, multiple, and confusing stories that have grown up around King Arthur's court and the Holy Grail. In the fifteenth century, Sir Thomas Malory brought gloriously back to Britain stories written down in France and Germany in the twelfth and thirteenth centuries, during the great era of the Gothic cathedrals. Those stories had originally been invented orally much earlier in Wales, Cornwall, and Ireland about events on British soil during its pagan past.* Over all these interlocking stories of Sir Lancelot and Guinevere, Sir Galahad, Perceval, and many others hovers an element of impenetrable obscurity. It can be explained in part by the intermingling of pagan ritual and Christian mysteries, and by confusions and changes in the transmission. Thanks to Tennyson's *Idylls of the King* and to Wagner's *Parsifal*, we have come to see in these stories the essence of the Middle Ages and of a Celtic mythology that sometimes rivals materials from both Greco-Roman and Judeo-Christian antiquity.

The entire Arthurian corpus can be read as a complex mystery story about knights who attain or fail to attain various forms of esoteric knowledge. After my two previous chapters on the perils of curiosity and presumption, I am duty-bound to take account of the establishing episode of the Grail story told both in Chrétien de Troyes' *Perceval* and in Wolfram von Eschenbach's *Parzival*. The episode assigns a different role to curiosity.

Having left his widowed mother and set out to seek knighthood and adventure, Perceval is directed by some fishermen to a strange castle, where the maimed lord welcomes him. A series of incidents—a grail that provides food for everyone, a bleeding lance, unidentified people hidden in adjoining rooms, and a magic sword—leave Perceval in profound puzzlement about where he is

---

*Roger Sherman Loomis offers a concise survey of the development of these materials in *The Grail: From Celtic Myth to Christian Symbol* (1963).

and what is going on. But his tutor knight trained him not to ask indiscreet questions, and he remains silent. It later turns out that Perceval's discretion has been his undoing, for he has missed the opportunity to ask the spell-breaking question that would cure the lord (the Fisher King, whose land is rendered sterile by his wound), avenge his father, and assure his own reputation.

Prometheus and Pandora, Eve and Adam, Psyche, and their ilk suffer dire consequences when they break a prohibition against seeking specified forms of knowledge. Perceval, heeding the warning he has been given against misplaced curiosity, fails the first great test of his manhood. By itself, the episode seems to favor a certain bold enterprise and even temerity in a knight. Set back amid the labyrinth of Arthurian stories, the Fisher King incident blends into an endlessly renewed quest for adventure and experience, forever out of range, never fully realized. Perceval plays a bumblingly human, almost comic role in an otherwise dark scenario. Lohengrin, Perceval's son, continues the quest for the Grail. Out of such materials, which include some famous love stories, was woven the fabric of medieval chivalry, an immense cultural excrescence on Christian doctrine.

The precariously balanced blend of ritual combat and hopeless love that makes up the ethos of chivalry provoked two dependent antichivalric stories that have grown into half myths. By steeping himself in chivalric lore, Don Quixote went harmlessly mad. His comic adventures provide the first stage in transforming the figure of the noble knight into a knavish picaro. The deep springs of Spanish literature also produced the other Don, who reduced chivalry to a tactic of unbridled and always unfulfilled egoism in the form of sexual conquest. Most versions treat Don Juan as a surprisingly sympathetic villain. Compared to the cautionary myths of the ancients, Arthurian romances with Don Quixote and Don Juan as outriders appear to encourage a growing boldness and independence of behavior in the face of traditional constraints. Did the hierarchical structure and closed intellectual universe of the Middle Ages lead to an existential impatience expressed in the new myth of chivalry? Such a surmise cannot be demonstrated. But the other myth of modern times seems to confirm such a view of how we shook off the Middle Ages.

Our second great modern myth without origins in antiquity concerns the restless middle-aged doctor-adventurer: Faust. Written versions of this legend do not reach so far back into the Middle Ages as those of the Grail legend. The story of the learned doctor who sells his soul to the devil in order to obtain supernatural powers shares with the chivalric tales a strong emphasis on the quest motif. Some scholars trace the learned-doctor theme back to Prometheus or to the powerful magician Simon Magus in Acts 8:9–24. But Faust's authentic origins lie in popular medieval stories and puppet plays about gaining knowledge from the devil. They seem to have converged on the historical figure Johann Faust, a scholar and charlatan in black magic who lived around 1500. But not until 1587 did Johann Spiess publish the first written version of the Faust story. In that chapbook, the learned doctor signs a pact in blood. He cedes his soul to Mephistopheles, the devil's messenger, at the end of twenty-four years, during which Mephistopheles "shall learn me [magic] and fulfill my desires in all things." Such a simple-minded plot indirectly expresses the Renaissance spirit of exploration as it moved north and the defiant spirit of the Protestant Reformation as it moved south.

For reasons not immediately apparent, all versions of the Faust story appear to be fragmentary and confused.* The powerful appeal of the situation never works itself out into a unified and convincing action. The story has attracted many writers; not even Goethe gave it a workable, definitive form. In Marlowe's earlier *Doctor Faustus* (1593), the character wants to be able to fly and become invisible, to be emperor of the world and a deity. A full complement of clowns, comic devils, and a Pope bamboozled by magic tricks turn the middle scenes into slapstick. The fifth act reduces Faust's final moments to moral allegory as stereotyped as *Pilgrim's Progress*. Weak-willed Faust wishes "I had never read book," and he has to listen to Mephistopheles' preachments: "Fools that must laugh on earth will weep in hell." Marlowe's still-medieval play stands closer to *Ubu Roi* than to high tragedy or to the anxieties of modern identity.

---

*Some of the confusion or ambiguity is carried in the name. In German, Faust means "fist," with conventional overtones of force, defiance, and ambition. The Latin Faustus means "the favored one," a form that can yield Prospero in English. It is instructive to read *The Tempest* as a modified Faust play.

After Marlowe came a spate of puppet plays in the marketplaces of Europe, exhibiting Faust catapulted at the end into the yawning jaws of hell. Audiences loved the lurid stage effects. It was the German dramatist Lessing, an unrelenting critic of French classicism and a champion of Shakespeare, who in the middle of the eighteenth century conceived the change that removed Faust from the Middle Ages and placed him squarely in the modern world. Though all but fragments of Lessing's Faust drama have been lost, we know that in his version Faust is not damned for his pact with the devil: He is saved.

That shift showed Goethe the way. Working in spurts throughout his lifetime, Goethe grafted Faust onto the Job story and produced a play so extended and episodic that the unity of dramatic action has been lost. It is rarely staged in a complete version; adaptations for opera amputate entire sections. When we reach the end of the play, we can attach only dubious moral and symbolic meaning to the fact that the sinner and playboy of the Western world is finally saved—because of his "striving." What, then, is Faust striving to achieve? In Goethe's version, as in earlier ones, we cannot readily find a scene in which Faust's nobility rises above his egoism. He has few redeeming qualities. In the newly introduced Gretchen episode, he is responsible for four homicides. The villain of the puppet plays has accomplished little to earn God's favor and final salvation.

I believe that we are drawn to this "tragedy," as Goethe called it, because it is chock-full of comedy. However, its publication in installments did not block the development of the legend in other directions by other authors. Halfway between *Faust I* (1808) and *Faust II* (1833), there appeared in London an anonymous novel called *Frankenstein; or, The Modern Prometheus* (1818). It soon lost its anonymity. In that remarkable book conceived when she was nineteen, Mary Shelley assimilated a wide range of classical and modern myths, from Prometheus to Milton's Satan to Locke's *tabula rasa*. Most importantly, she takes aim at the Faustian motif of "the serpent sting" of knowledge. There are many reasons to read these two books together.

2. Two Conflicting Versions

The twenty-five scenes of Goethe's *Faust I*, without further division into acts or sections, fall roughly into three sequences: abdication and changed allegiance; seduction of Gretchen and betrayal; flight and remorse. In the late midlife crisis of the opening scene, Faust puts aside all his attachments—book learning, language itself as a path to knowledge, his high status in the community, his links to the institution of the university—in order to do a deal with the Devil's agent. Having cursed everything from fame to family, from money to faith, he seeks and fleetingly finds pure pleasure, the rush of experience for experience's sake. To Gretchen's question about his religious beliefs, Faust has a revealing answer.

Fill your heart to overflowing,
and when you feel profoundest bliss
then call it what you will:
Good fortune! Heart! Love! or God!
I have no name for it!
Feeling is all;
the name is sound and smoke,
beclouding Heaven's glow.

(3451–58; tr. Peter Salm)

This modern Job figure is willing to call his sensuous bliss his God, a clear declaration of hedonism. In the biblical Job, such blasphemy would have immediately removed God's favor; in Goethe's play, Gretchen observes mildly that there's something awry in his confession, and the scene moves on. The innocent-seeming Gretchen romance, punctuated with delicately lyrical moments, leads to a succession of disasters from which Faust walks away—or flies away when Gretchen is saved after her death. Bliss and feeling overcome all scruples.

Faust's rejection of conventional rewards in order to seek for the intensity of experience is framed in a series of three portals through which one enters the work. The dedication in effect recommits the book to Goethe's own youthful imagination, whose spirit world he comes upon twenty years later in manuscripts he put aside. The "Prelude in the Theater" insistently tells the reader or spectator through the nonclassical personage of the Clown *(Lustige Person)* to expect a complex mixture of truth and error.

> *We must present a drama of this type!*
> *Reach for the fullness of a human life!*
> *We all live it, but few live knowingly;*
> *if you but touch it, it will fascinate.*
>
> (166–69; TR. PETER SALM, MODIFIED)

After this manifesto of a working theater director, the "Prologue in Heaven" descends abruptly from the Archangels' lofty celebration of the Lord's created universe (243–70) into a jocular exchange between the jester-trickster Mephistopheles and the enormously tolerant Lord himself. The Lord even welcomes Mephistopheles' impertinent bet that he can corrupt Faust, for the Lord states that it may take a rogue *(Schalk)* to goad human beings out of their apathy. Every critic from Schiller on down has had to deal with the enormous shifts of tone and mood in the play. Goethe himself spoke of "serious jests."

Should we take *Faust I* seriously? Mephistopheles' constant jocularity keeps us guessing. And the "Prologue in Heaven" initiates an elaborate metaphysical riddle, bordering on a joke, adapted from the Old Testament. *Job*: Why do the godly suffer? *Faust*: Why are the ungodly saved?

It is difficult to say how far the hedonism of *Faust* reflects Goethe's life and times. In this extensive work, his genius rises easily above ready-made categories like classic and romantic, science and poetry, spiritual and demonic, social and individual, tragic and comic. At the tightly organized Weimar court, Goethe committed himself to statecraft, to running a theater, to scientific research, and to a substantial array of friends and admirers. In comparison, the character Faust looks like a loner lost in unfamiliar territory. As the French Revolution engulfed Europe in turmoil, Goethe seemed to move toward more lofty accomplishments. But Goethe, the unchallenged founder of modern German literature, stayed loyal through thick and thin to this jagged play about estrangement and dissatisfaction with life. It would not let him go. Yet Faust, the striver and overreacher who is spared his punishment, remains in great part a literary and cultural enigma.

On the other hand, the circumstances of Mary Shelley's life offer clear pointers about why she wrote her first novel, and how she could finish it in a year at such a young age. She lived her earliest years with famous people admired by many for their genius, their high ideals, and their presumably rewarding lives. But her widowed father, William Godwin, was a notorious socialist whose utilitarian morality induced him to write that in a fire he would save a treasured book before a member of his own family. He hardly knew how to take care of his daughter. She knew her mother, Mary Wollstonecraft, who died in childbirth, only by the stories of her dedication to feminism, revolutionary causes, and friends in need. Percy Bysshe Shelley, the stereotype of the Romantic poet, carried Mary off at seventeen to the Continent without marrying her, to live for a time in the irregular household of another Romantic poet, Lord Byron. Surrounded by illegitimate births and infant deaths, they subsisted on high ideals to remake the world through liberation and revolution. The men in the group were intent upon achieving glory through their genius; other concerns must not stand in their way. Still in her teens, Mary surrendered a part of her being to this heady life, for which the rest of the world might well envy her. She was the ultimate Romantic groupie. But she also perceived so vividly the vanity and selfishness of this existence that she produced a narrative account of it already halfway to myth. One may well find *Frankenstein* in many passages an ill-written and exaggerated novel. But its remarkable narrative structure holds in place a story whose pertinence to the history of Western civilization has grown from the day it appeared. Whereas *Faust* has the appeal of an eternal enigma, *Frankenstein* has the sting of a slap in the face to the author's own kith and kin.

Frankenstein deploys an array of machinery as complex as *Faust*'s to draw us into its story. The subtitle makes a hugely ambitious claim by presenting the novel's hero as "the Modern Prometheus." The epigraph rings in a stark quotation from Adam in *Paradise Lost* to describe the abandonment felt by the creature whom Dr. Frankenstein galvanizes horribly into life.* In the original anonymous edition of 1818, the dedication to William Godwin, which led many to believe that Percy Bysshe Shelley had written the book, was followed by an unsigned preface, which Percy did write for Mary. "I have endeavoured to preserve the truth of the elementary principles of human nature, while I have not scrupled to innovate upon their combinations." Writing in the person of Mary, Percy is affirming the exploratory side of the story, presented as an experiment in human nature that observes, like Poe's stories and modern science fiction, basic psychological principles. Then a set of letters by Walton to his sister in England describes both his own expedition toward the North Pole and his encounter in the Arctic waste with Frankenstein, a fellow scientist in pursuit of glory through great enterprise. Finally, the exhausted Frankenstein narrates to Walton his lengthy story of creating a living monster out of cadavers. At the center, embedded in Frankenstein's tale, one comes upon the monster's story, told on a spectacular glacier high in the Alps. The effect of all this narrative nesting is to ensure that the mother story is taken in dead earnest. This godless universe, provided nevertheless with spirits and demons and all the elevating effects of the sublime in nature, provokes not a single intentional smile or laugh to attenuate the murders of four people close to Frankenstein by his own creature.

Let me restate the two actions. Having achieved high social and intellectual status in life, Faust abandons it for doubtful accomplishments as romantic lover and fantasy traveler. Across three continents, he practices impatience with himself, with Mephistopheles, with all creation. Young and unknown, Frankenstein seeks fame, the only salvation offered in his faithless world. He throws himself into the fanatic attempt to create human life, an act traditionally limited to a god figure. By succeeding, he damns himself. Frankenstein also is responsible for four homicides. "Learn from me,"

---

*"*Did I request thee, Maker, from my clay*
To mould me man? Did I solicit thee
From darkness to promote me?—"

(PARADISE LOST, X, 743–45)

he tells Walton, "how dangerous is the acquirement of knowledge, and how much happier that man is who believes his native town to be the world, than he who aspires to become greater than his nature will allow." But Frankenstein hardly means what he says.

Despite the differences in dramatic outcome and in pervading tone, these two tales of metaphysical adventure turn out to be the most effective and lasting versions of a single myth: the learned doctor discontent with his lot and seeking release into superhuman life.

3. Scenes from *Faust*

To a remarkable degree, the opening scenes in Faust's study recapitulate the first two parts of Descartes' *Discourse on Method.* Descartes tells us how he abandoned the study of literature, mathematics, theology, philosophy, law, medicine, and rhetoric for more practical knowledge to be gained from travel, experience, and common sense. Faust tells us that he has an advanced degree in all those fields. The difference between the two stories lies in their timing, in where they pick up the thread of the action. We come upon Faust in his study just when he is impatiently trying to break out of his musty learning in order to seek a life of action. We come upon Descartes just as he settles back into his study *(poêle)* after years of soldiering and travel. What Descartes describes as being behind him forms not a bad summary of what still lies ahead of Faust. Three hundred years later, these sentences retain a trenchant timeliness.

> *I completely abandoned the study of literature. Deciding to seek only that knowledge I could find in myself or in the great book of the world, I devoted the rest of my youth to travel, to visiting foreign courts and armies, to frequenting people of diverse characters and conditions, to accumulating varied experiences, to testing myself in whatever encounters came my way, and at all times to reflecting profitably on these events. For it seemed to me that I would discover much more truth in the reasonings of men about what they know directly, men who will bear the consequences if they make a bad*

decision, than in the reasonings of a scholar in his study, who produces speculations without application and without consequence to him, except perhaps the vanity he finds in their remoteness from common sense. . . .

(*DISCOURSE ON METHOD*, Part One)

Descartes could be speaking for Faust at the opening of Goethe's play. Then, with Mephistopheles as tour guide and tutor, Faust flies off to seek the practical knowledge and experience of the ways of the world from which he has sheltered himself. Unlike Descartes, Faust never returns to his study to take stock of what he has learned. His experiences and enterprises go on and on. Death alone can close the structure of the play.

Any museumgoer knows that a common subject in Renaissance painting is Saint Jerome in his study. He is depicted in his monastic cell, with books, cross, and death's head. Like Marlowe, Goethe chose Faust's study as the principal scene for his intellectual drama, to which the Gretchen story forms an awkward yet appealing appendage. Having dismissed all traditional fields of study in the first scene and invoked any nearby spirits in the second scene, outdoors, Faust discovers that a spirit (in the form of a poodle) has followed him back into his study. After comic conjurations, Mephistopheles stands before him "dressed as a travelling scholar"—that is, as Faust's parodic double. Faust is the one to propose "a pact," as if he already knew the particulars of his own myth from earlier sources. Mephistopheles stalls; his attendant spirits put Faust to sleep so that this lesser Lucifer can consult with higher authority.

When Mephistopheles returns, Faust is in a foul mood and curses "all the things that now entice my soul" (1587). The curse includes the very faculty of imagination: "The god that lives within my bosom" (1566) and that drives him away from dusty books to seek the sublime. All the discussion here is both very abstract (unless convincingly staged) and improbable as a prelude to the big moment. It takes a spirit chorus to talk Faust back down to tractability so that Mephistopheles can deal with him. By declining any conventional offer of gold, girls, and glory (1679–87) Faust rejects the historical quid pro quo of a soul exchanged for a period of magical bliss. Instead, Faust proposes a wager.

If ever I should tell the moment:
Oh, stay! You are so beautiful!
Then you may cast me into chains,
then shall I smile upon perdition!

(1699–1702)

Thus the traditional contract, which gave Faust nothing to do but to enjoy himself for twenty-four years, is changed into a competition to see who is the wilier.* A wager leaves Faust the possibility of winning, of having it both ways: both exploiting Mephistopheles' supernatural powers and gaining final salvation following Lessing's version.

It is important to note that before the "end," far distant in both Faust's and Goethe's lives, Faust has essentially lost his wager at least twice. In the "Martha's Garden" scene, he contemplates his love for Gretchen as inexpressible.

. . . to give oneself completely and to feel
an ecstasy which must be everlasting!
Everlasting!—for the end would be despair.
No—no end! no end!

(3191–94)

This would appear to be the *Augenblick* ("moment") snatched out of *das Rauschen der Zeit* ("the rush of time," "the stream of consciousness," [1754]), the moment of bliss to which Faust has wagered he will never submit completely. In Part Two he surrenders in similar ecstatic fashion to Helen (9381–418). But somehow the march of events brushes by the wager that started the action. Neither Mephistopheles nor the Lord ever calls Faust on the bet he has lost. Thus Goethe collapses the Job story into a fiasco saved at the end only by a miracle.

*In the Middle Ages and Renaissance, a guarantee of twenty-four more years to a mature man represented a substantial gift of longevity. Christopher Ricks has pointed out the importance of this element to Marlowe's Faust. By 1800, statistics and circumstances had probably changed enough to make life expectancy a less compelling consideration for Goethe's hero.

All editors identify the book of Job as the source of Mephistopheles' wager with the Lord. Too few editions point out that we also know where Goethe found the idea for the second wager.* In the fifth section of *Reveries of a Solitary Walker*, Rousseau evokes his idyllic life of solitude and idle meditation, of *dolce far niente*, on the Island of St. Pierre in a Swiss lake. Adrift in a skiff on the calm water, he accomplished no exploits, earned no glory. Instead, by a beautifully described process of renunciation, he attained "the feeling of existing at the simplest level." It soon becomes the most exalted level. Rousseau's reflections on this state of being mark an important and troubling moment in the spiritual history of the West.

> *Thus our earthly joys are almost without exception the creatures of a moment; I doubt whether any of us knows the meaning of lasting happiness. Even in our keenest pleasures there is scarcely a single moment of which the heart could truthfully say: "Would that this moment could last forever!" And how can we give the name of happiness to a fleeting state which leaves our hearts still empty and anxious, either regretting something that is past or desiring something that is yet to come?*
>
> (TR. PETER FRANCE, 89)

This yearning to surmount the flux of time and to eternalize the moment contains both a mystical and a blasphemous element. Rousseau acknowledges his hubris a few lines later: "What is the source of our happiness in such a state? Nothing external to us, nothing apart from ourselves and our own existence; as long as this state lasts we are self-sufficient like God" (90).

Goethe responded to Rousseau's aspirations to transcendence by having Faust refuse (with two exceptions) temptations to transcend time. He does not, as in Marlowe's version, sell his soul for two guaranteed decades of high living. He wagers that no feeling, no matter how profound, that no human attachment will ever lure him into loyalty. That stony-hearted principle allows Faust to try anything a few times, like an intellectual philanderer or a participant

---

*A good discussion appears in Chapter Four of Jane K. Brown, *Goethe's Faust.*

in a sexology research experiment. He always moves on. Nothing is at stake beyond his own opulent survival.*

The moral of Faust's life and of Goethe's drama cannot be easily grasped. It lies deep in paradox and ambiguity. Faust clings to contingency yet wishes to rise above it. "Striving" looks both toward high aspirations and toward irresponsible opportunism. Faust covets divine status. By turning down Mephistopheles' usual blandishments and by insisting on an open-ended deal that gives him Mephisto's magic powers for as long as he remains unsatisfied, Faust tricks both Mephistopheles and the Lord into granting him higher status than mere mortality. "Oh, if I had wings," cries Faust in his prophetic "Sunset" Speech. Three scenes later, he is flying all over Europe and enjoying his "godlike course" (1081).

Even before *Faust I* was published in 1808, it was declared a masterpiece, the culminating work of Europe's most celebrated man of letters. The unplayable play seemed to subsume and surmount the social and artistic conflicts of that revolutionary era. Since Goethe continued working on it intermittently for two decades until his death, the unfinished play enjoyed the status of a monument in progress of world literature encompassing Romantic and classic impulses. In our time, a company of devoted actors performs the entire drama every few years at the Steiner Institute in the Swiss town of Dornach. The ritual takes several days. College students in many countries read Part I attentively. Several operas have drawn their scenario primarily from the Gretchen episode, Goethe's addition to the original story. The adjective *Faustian* has passed into many languages.

Goethe's *Faust* deserves its many honors on two grounds. First, Goethe identified one of the great dramatic situations afflicting and driving human beings in the modern world. We strive without knowing adequately what we are striving for and we believe our

---

*So described, Faust's attitude of self-gratification resembles that of many characters in the novels of a French author writing during the same revolutionary period. One could read the heinous episodes of the Marquis de Sade's *Juliette* as a violently dehumanized caricature of Faust. Having made a semiwager to outshine and outperform her virtuous sister, Justine, Juliette conquers Europe by abandoning all constrainsts, all scruples, and all feelings. And the gods favor her triumph by destroying her victimized sister with a symbolic bolt of lightning. I shall deal further with Sade in Chapter VII.

thirst for knowledge and experience is protected in high places. Apparently, the hunch about the Faust story came to Goethe as a twenty-four-year-old law student in Strasbourg. We do not know when he decided on the two major changes that transformed the archaic medieval plot of magic into a modern psychophilosophical myth—namely, substituting an open wager for the twenty-four-year pact, and substituting salvation for damnation.

Second, Goethe poured out of himself a river of masterful German poetry in a variety of moods and verse forms. No major work of literature by a single hand attempts to mix so many different styles, a virtuoso accomplishment that has the consequence of rendering adequate translation close to impossible. The "Sunset" Speech (1064 ff.) builds into a full-throated Romantic ode to flight. Gretchen's song while undressing in her bedroom has passed into folklore like Shakespeare's songs. Here German and English come very close.

Es war ein König in Thule
Gar treu bis an das Grab,
Dem sterbend seine Buhle
Einen goldnen Becher gab.

There was a king in Thule,
Was faithful to the grave.
To him his dying lady
A golden goblet gave.

(2759–62, TRANSLATION MODIFIED)

Faust and Mephistopheles joust and mock one another constantly in the popular, freely varying *Knittelvers* of archaic puppet plays. Compared to *Paradise Lost*, even considering the remarkable mood changes Milton could inject into his ten-syllable line, Goethe's twelve-thousand-line drama reads like a poetic variety show or a three-ring circus.

A powerful situation and dazzling verse demand our attention and our admiration. Nevertheless, as a play, as an episodic tale of a larger-than-life hero, *Faust* does not fulfill either Goethe's expectations or ours. Faust scholarship has been loyal and enormously resourceful in interpreting the work. But for all its remarkable scenes and entertaining moments, *Faust* lacks the one unity we continue to look for: unity of action. Life, of course, does not usually happen to us in neat units called "actions," nor can we make it happen that way. But we seem to yearn for that coherent shaping of experience. In short conversational anecdotes, in great oral epics, and in the intensified timing of a short story, we have created for ourselves a sense of narrative movement and moral significance that has a discernible completeness of shape on the scale of human events. No culture has been discovered without its storytellers to record and recapitulate the life of the tribe. As complementary evidence of our yearning for coherent stories, all cultures have also produced some form of the cock-and-bull story, a nonsense version of events that improvises incidents without shape or direction. Such sheer contingency makes us laugh. Seeking originality, some modern and "postmodern" authors have turned toward this formlessness.

But even in Part Two, *Faust* is no cock-and-bull story. Goethe's immense play aspires to a unity it does not attain. By default, therefore, the play can be seen as belonging to several modern categories—theater of the absurd, cinematic montage, and compulsive self-parody. These aspects of the play point forward toward Ibsen's *Peer Gynt* and Jarry's *Ubu Roi.* But we should not stray too far from Goethe's central project. The greatness of *Faust* lies more in its theme—human greatness contains human weakness—and in its dazzling poetry than in the way Goethe assembles its many parts.

Writing in 1795, when only fragments of Goethe's *Urfaust* had appeared, Friedrich von Schlegel praised the magnificence of the poetry and the "truth" of its philosophic content. Schlegel felt no qualms, even on fairly slender evidence, about comparing Goethe to Shakespeare. "Indeed, if *Faust* were to be completed, it would probably far surpass *Hamlet* . . . with which it seems to have a common purpose." To which I would respond that Goethe never really did complete his drama; he just kept adding to it. And if the "purpose" it shares with *Hamlet* concerns the difficult passage from thought into action, neither the wager motif nor Faust's ultimate salvation genuinely illuminates it. From the start, Goethe produced a monument already in magnificent ruins, a modern Sphinx or Acropolis, a drama in progress for a lifetime and one that had to weather the constant buffeting of its creator's imagination. Born a classic, *Faust* comes to life in flashes, not as a whole.

But among stories of forbidden knowledge, *Faust* looms very large. In creating his modern hero, Goethe stands Adam on his head. Faust seeks knowledge beyond all bounds, beyond his *portée.* He breaks the Christian taboo on pagan magic. He scorns Descartes' judicious return to his study after gaining adequate experience of the world. And then Goethe asks us to believe that this privileged, self-indulgent scholar, not misled by the blandishments of any scheming Eve, should be forgiven, even praised, for his "striving." Here is our modern Adam, raised up to heaven by a chorus of angels for conduct more proud and defiant than what earned the original Adam banishment from Paradise.

Milton handled things differently. In an epic yet often down-to-earth retelling, he foresaw Adam's redemption through the Fortunate Fall without suspending his judgment or his punishment. Truth here has its consequences. Goethe, on the other hand, never frets about disobedience. He calmly usurps the Lord's role and reverses the verdict, quashes the sentence on his new Adam. Now the truth need have no consequences. For Faust, all is pardoned in advance.

Mary Wollstonecraft Shelley, writing soon after *Faust I* appeared, rejected both Milton's Adam and Goethe's Adam. She imagined not only a new Adam as creature-monster driven to despair and depravity but also the Promethean hubris that led to his creation not by a god but by a presumptuous mortal. It is hard not to read her novel as a retort to *Faust.*

4. Scenes from *Frankenstein*

In an early episode of *Faust II,* Mephistopheles wanders into the laboratory of Wagner, Faust's former graduate assistant, now an advanced research scientist in genetics. At that very moment, Wagner succeeds in creating in a luminous, vibrating alembic the entity Homunculus, pure humanoid mind without a material body.

Goethe treats the miraculous incident as pure self-parody—a miniature, disembodied Faust in a bottle seeking full being and mouthing such pseudo-Faustian lines as "Since I exist, I must be ever active" (6888). Homunculus calls Wagner "Papa" and Mephistopheles "Sir Cousin" and spies on Faust's erotic Leda dream. As if to underline the jokey aspect of the sequence, Goethe later suggested in a conversation with Eckermann (December 30, 1828) that Homunculus would make a good part for a ventriloquist. In *Faust II*, jest occupies far more surface area than earnest.

Written a decade earlier than Wagner's dabbling in genetic experiments, *Frankenstein* never jests and never forgets that the artificial production of life carries dire consequences. Immediately after Frankenstein has animated the "creature," the enterprise is given the epithets "catastrophe . . . horror," an operation bringing into being a "wretch . . . monster . . . daemonical corpse" (Chapter 5). Frankenstein flees to his bedchamber and dreams of Elizabeth, his foster sister and true love. In his embrace, she turns into the corpse of his dead mother, crawling with maggots. It is hard to avoid a symbolic interpretation: Frankenstein, hoping to achieve a scientific miracle deserving admiration, discovers that he has violated Mother Nature herself.

Goethe treats the creation of new life as an incidental joke; Shelley places it at the center of her story and sees it as a monstrous aberration. The contrast can be explained only in part by the differing lives and temperaments of an indulgent, aging survivor of both the Enlightenment and Romanticism and of a bookish young girl not duped by the men whose genius she admired. Goethe's comic incident would have revealed a tragic side to a teenage mother whose first child died eleven days after birth.

The incidents of Shelley's novel build inexorably toward the climax of intellectual ambition unmasked. It provides her grand finale. The all-too-human monster, who has tried earnestly, though implausibly, to socialize and educate himself, commits four horrible murders among those Frankenstein loves most. The monster flees into the Arctic wastes, pursued by Frankenstein. The action devolves into a grotesque contest in madness, self-glorification, and self-immolation. The dying Frankenstein shows great agitation as he speaks to Walton, the fanatic explorer who is trying to rescue him. "Farewell, Walton! Seek happiness in tranquility and avoid ambition, even if it be only the apparently innocent one of distinguishing yourself in science and discoveries. Yet why do I say this? I have myself been blasted in these hopes, yet another may succeed" (Chapter 24). The self-challenging question and reversal of direction toward the end of the passage require a distinct pause and mark the reappearance of the fanatic scientist wanting to pass the torch.* Even in death, Dr. Frankenstein, the Modern Prometheus, cannot lay aside the ambitious drives that have devastated his life.

Enter now the demon, or monster. In the four closing pages, he delivers a harangue to Walton over Frankenstein's corpse. The monster claims melodramatically to have suffered even more than Frankenstein, who lost all his dear ones by violent murder. "My agony was still superior." The demon will assemble an immense funeral pile on which to be consumed "triumphantly." His apotheosis is as grotesque as it is melodramatic. The battle to which these awful adversaries commit themselves is the struggle for glory, the driving male condition that inspired Mary Shelley to write the book in horror and in protest. The monster usurps the role of suffering Prometheus from the man who created him. Little wonder that in the resulting myth and in popular parlance, the name Frankenstein is often transferred from creator to creature.

5. Related Stories

Time sometimes reverses itself. The best spoof of *Faust* preceded it in the history of European literature rather than followed it. The other great anti-intellectual hero spent so much time pouring over books of chivalric lore that he was driven simultaneously insane and out into the world in quest of high adventures. Here is a light-

---

*Stephen Jay Gould has recently argued that Dr. Frankenstein's motivations as a scientist "are entirely idealistic" but that he failed to "undertake the duty of any creator or parent" to assume responsibility for his offspring. The second proposition is unimpeachable. In making the first, Gould fails to perceive how carefully Shelley describes Frankenstein's brief moment of idealism (Chapter 4) yielding to the "frantic impulse" of hubris and egoism.

hearted version of forbidden knowledge. This learned doctor decided to become a knight. It takes Cervantes one short chapter to launch Don Quixote de la Mancha into the domain of realities crossed with fantasies. The fanfares and negotiations surrounding Faust's setting forth consume ten times the space. As soon as he gets out on the road, Don Quixote starts talking to himself and lets his nag Rocinante choose their path toward adventure. "Undoubtedly in days to come when the true history of my famous deeds *[hechos]* comes to light . . ." he muses. Cervantes has us laughing from the beginning over the preposterous exploits of the scholar turned adventurer.

That endlessly extensible episodic situation based on the conventions of chivalry anticipates the scene of Faust in his study, where he opens the New Testament to translate John 1:1. For *logos*, he brushes aside successively *word, mind,* and *power* in order to settle on Don Quixote's *hecho*—in German, *die Tat;* in English, *deed.* All three are substantives based on the infinitive *to do.* Had *Don Quixote* appeared after *Faust,* the literally crazy exploits of the knight of La Mancha would have been interpreted as a superb send-up of Faust's carryings-on with Gretchen and later with legendary figures from all history. Don Quixote starts out alone on his quest and is sometimes reduced to talking to himself and to reciting stories remembered from his books of chivalric lore. Only in Chapter Seven does he persuade a "hapless rustic" to become his squire by promising him an island to govern. Thus Sancho Panza fills the role of traveling companion, confidant, and remonstrator satirically symmetrical to that of Mephistopheles for Faust in his travels.

The alert reader will already have glimpsed another pair of elegantly disreputable characters lurking in the neighborhood—Don Juan and his scalawag servant comb the landscape not for damsels in despair needing a knight's help, but for any woman vulnerable to a man's advances. No aura of dusty book learning clings to Don Juan. He is a man of duels and trysts and pursuits. But beyond that, he seems to elude our grasp by dodging in and out among the several masterpieces that have brought him to life. In Tirso de Molina's original *El Burlador de Sevilla* (1630), Don Juan is a madcap deceiver whose principal pleasure comes from having tricked one more woman (and usually one more husband) and whose defiance of convention does not arise from loss of religious faith. The original Spanish drama with the stone statue of the Commander calling down God's wrath remains close to an *auto sacramental,* or miracle play.

Molière's *Dom Juan* (1665) presents a highly sophisticated modern scoundrel who requires a constant change of female diet to defeat boredom and whose pleasure is not in tricks but in metaphysical conquest. Da Ponte and Mozart simplified the story and gave increased importance to the female roles in *Don Giovanni* (1787). They produced not grand opera and not high tragedy, but a *dramma giocoso,* which portions out both tricks and miracles. Should we refer to Don Juan as Faust without university degrees? What does it mean that our Western literary tradition has selected these two selfish opportunists to celebrate in a series of major works? Where then does Don Quixote fit into the procession? What form of greatness of character or of moral vision is offered to us in these works?

The dissatisfied German doctor who deludes himself that he wants a life of action will never displace the nutty knight who truly loves and lives by his books of chivalry, or the irritable self-defeating Spanish womanizer. Still, there is one more common feature worth pointing out. All three figures are closely accompanied by a companion and foil whose role is both to serve and to mock. Like Plato's dialogues that flicker with Socratic irony and Proust's novel that sustains the no-nonsense crankiness of the servant Françoise through three thousand pages, these three stories embody their own parody and criticism. That fact represents a partial answer to the questions asked at the end of the preceding paragraph. Mephistopheles punctures Faust's bubbles of pride and Romantic sentimentality soon after they form, and in a few scenes he outshines his famous rival in the great wager. The two servants representing ordinary common sense for the two Dons become almost as bold as Mephistopheles. Each of these works provokes frequent laughter at the expense of its hero's extravagant ambition.

In contrast, *Frankenstein* offers not a single comic moment. The story's Romantic excesses, as in Safie's abduction story and the funeral pyre competition at the end, provoke impatience in the reader rather than guffaws. For all the complicated narrative through letters, transcribed stories, and stories within stories, Mary Shelley never makes a move to undermine the high seriousness of her bloodcurdling tale. Byron and Percy Shelley took it as something of a lark during the trying summer of 1816 in Switzerland to have a go at writing ghost stories. Mary remained stern and unyielding. Her judgment of the presumptuous and selfish actions of Frankenstein in creating and then abandoning a new form of life is nowhere softened in the novel. She minces no words to tell us that for all his striving, her Modern Prometheus deserves not the glory he seeks but the humiliating death he finds in the barren wastes of the Arctic.

The resolute moral stance of *Frankenstein* about observing our human limits can be seen now as exceptional. Other great modern works were proposing a relaxation of both classic and Christian moral traditions. Milton depicted the Garden of Eden as the scene not of a tragedy but of a Fortunate Fall. First Lessing and then Goethe transformed the figure of Faust from greedy charlatan into transcendental hero, linking the Enlightenment to Romanticism. This gradual attenuation of guilt also affects the story of Don Juan. In early versions, the stone statue of the Commander sends the unrepentant sinner to the tortures of Hell. When Romantics like Hoffmann and Grabbe and Kierkegaard got their hands on him, the Spanish lady-killer was recostumed for moral rehabilitation. Théophile Gautier made the simplest case by calling him the "Faust of love." Elsewhere, Gautier explained: "Don Juan goes not to Hell but to Paradise, for he sought true love." Salvation came flowing in from all sides, even if it meant rewriting the story and tidying up the leading man. The Romantics often did not seek harsh judgment of their scoundrel heroes.

Apparently, it required a woman to inventory the destruction caused by the quest for knowledge and glory carried to excess, and to invent the counterplot to *Faust.* The Lord does not intervene to save Frankenstein; Mary Shelley's judgment is keener and more courageous than Goethe's cosmic leniency. Born and raised in the most notorious literary household of her day and believing that she embodied the spiritual heritage of Juliet and Desdemona, Mary Shelley threw herself at seventeen into a histrionic life surrounded by poets and geniuses. Three years later, in her first book, she was able to assess with lucid severity the compulsions of fame and glory that drove her companions and infected her. We have not yet exhausted her remarkable fiction that flies in the face of the Romantic and utopian themes that spawned it. Through its complex structure of narrative frames and embedded stories, *Frankenstein* maintains a sturdy-enough unity of purpose and action to give an ironic twist to the constantly invoked words *glory* and *honour.* Ten pages before the end, Walton says of the dying Frankenstein, "He seems to feel his own worth and the greatness of his fall." By this time, we know how much salt to add. Shelley has not deployed any battalions of angels to carry him off. This is no Fortunate Fall. No one can redeem the destruction Frankenstein has left behind him.

The numerous progeny of these two matching stories about wanting to know too much tells us that the motif of forbidden knowledge remains with us in multiple forms. *Faust* and *Frankenstein* together appear to have spawned a line of tales about doubles, Doppelgängers, locked in a struggle to destroy each other. Poe's "William Wilson" (1839) prepares the way for R. L. Stevenson's *The Strange Case of Dr. Jekyll and Mr. Hyde* (1886) and for Oscar Wilde's *The Picture of Dorian Gray* (1891). The three tales carry a strong dose of horror because they have turned the Faust-Frankenstein story inward. The protagonists summon an evil spirit not out of the surrounding environment but from inside themselves. A repressed portion of their character haunts them. Thus they come to know too much about their hidden being and can no longer believe in their own integrity. They can only squirm. None is saved.

Throughout the nineteenth and twentieth centuries, similar stories of forbidden knowledge proliferate faster than I can track them. Hawthorne centered two of his most obsessive tales on closely related themes. In "The Birthmark," a fanatic scientist discovers how to remove the tiny flaw in his wife's ideal beauty—and thus kills her. Ethan Brand, in the story that bears his name, seeks to know the unpardonable sin. He finds it less in the fiendish but undisclosed "psychological experiment" he carries out on a girl whose soul is destroyed than in the intellectual pride of his enterprise and within his own heart. The theme of destructive knowledge crops up again in "Rappaccini's Daughter," in *The Blithdale Romance* (1852), and in practically everything Hawthorne wrote. Thomas Mann tried to get a whole new grip on the Faust legend through the demonic forces of music and sexual thralldom in *Doctor Faustus* (1947). There is no slackening in our own anguished times. I detect a powerful Faustian strain in one of the most ambitious of

Woody Allen's films, *Crimes and Misdemeanors* (1989). Having learned from his Machiavellian or Mephistophelian brother that the woman threatening to ruin his life can simply be rubbed out, a successful eye doctor cannot resist the temptation to act on that knowledge. Ultimately, he is neither saved nor damned. He survives his guilty feelings and speaks at the end in telltale Faustian terms of the need to "keep trying."

The many Hollywood sequels to *Frankenstein* have manipulated the man-made monster situation in ways that cast the scientist in a particularly unfavorable light. All written and filmed works in the immense category of science fiction have their roots in the ground prepared by *Faust* and *Frankenstein* with their opposing attitudes toward forbidden knowledge. Those two stories will stay with us for a long time.

6. FAUSTIAN MAN: THE PRINCIPLE OF EXCESS

The term *Faustian man* has been accepted in English and several other languages in large part because of the German philosopher Oswald Spengler. He used the expression in his widely read *The Decline of the West* (1918), which develops a cyclic view of history. What is the social and moral content of the expression? On that point, Spengler is not our best authority. Let us look again at the opening and closing episodes of Goethe's version, both written during an intense period of work on the drama around 1800. "The Prologue in Heaven" enacts the inaugural wager for Faust's soul between the Lord and Mephistopheles; presumably the whole action hangs from this affirmation of faith in Faust's perpetual seeking beyond any human satisfaction, and from his later complementary bet with Mephistopheles to the same effect. "Midnight" and "Outer Precinct," the last scenes of Part II in which Faust appears alive, show us an aging, greedy empire builder irritated that his land-grabbing has killed three innocent victims. In his angry discussion with the crone, Gray Care, Faust makes two crucial and interlocking claims. First, he has lowered his sights from his earlier transcendent aspirations to godhead.

I know full well the earthly sphere of men—
The yonder view is blocked to mortal ken.

(11441–42; TR. WALTER ARNDT)

Second, his striving will not cease but will restrict itself to "this planet's face" (11449). Even though Care then blinds him, Faust is determined to carry on his settlements. Marshall Berman calls *Faust* "the tragedy of development" and links the modernizing schemes of Part II to Hitler's and Stalin's social engineering projects.

In the following scene, speaking rhapsodically of founding a City for free people and repeating word for word his wager with Mephistopheles about his never wanting a moment to last, Faust dies with the word *Augenblick* ("moment") on his lips. Death becomes his ultimate fulfillment, the satisfying moment he wishes to render eternal as his apotheosis.

Immediately, Mephistopheles responds by claiming that he has won both wagers. He has ample grounds. Faust's self-satisfaction in dying has betrayed him. Readers will also remember that Faust bears responsibility for seven homicides along the way. But a brief final interval of blindness is all the punishment he will receive. Goethe and the Lord have long since decided to save Faust, and the necessary machinery is largely in place. In the "Entombment" scene, while a chorus line of handsome angels distracts Mephistopheles, other angels carry off Faust's immortal essence. In this deliberately grotesque scene of score settling and soul snatching, Mephistopheles' outburst is entirely justified.

I have been robbed of costly, peerless profit,
The lofty soul pledged me by solemn forfeit,
They've spirited it slyly from my writ.

(11829–31)

A Christian deus ex machina cheats the devil of his due from two formal bets. It would be hard to contrive a more arbitrary and unearned ending to the lengthy drama.

There may be a precedent to help us grasp Goethe's thinking.

Cain, who murdered his brother and went on to build the first city, was cursed by the Lord and then granted protection from vengeance by others. The Lord needed Cain in his role as founder of civilization. In his last speeches, Faust sounds like a megalomaniac Cain. The angels bearing Faust's immortal essence sing about "striving" as the justification for his redemption, and we know from Goethe's conversations with Eckermann that he took this argument very seriously. But in the play, Faust capitulates three times to the spell of the moment and stops striving: with Gretchen (3191–93), with Helen (9381–82), and in his own vainglorious death (11581–86). Does Faust deserve salvation in spite of the wrecked lives he has left behind him? Should we even raise the question in the face of claims about "striving" and (in the closing lines of the play) about the Eternal Feminine drawing us upward?

A dispassionate survey of Faust's behavior would justify our protesting that the mawkish allegorical goings-on in the last scenes merely distract us from Faust's malicious, selfish, and sometimes criminal conduct. He has not attained spiritual regeneration. He lowers his sights from transcendent to mundane goals near the end and then reaffirms his megalomania. A curious case, all in all, approaching the world turned upside down. Evil, when associated with striving, turns into good. Is this the crowning work of the Enlightenment? Or of Romanticism? In one of the earliest intelligent responses to the already-enshrined masterpiece, Mme de Staël observed in 1810 that Goethe had created a story of "intellectual chaos" in which the devil is the hero and which produces "the sensation of vertigo" (*De l'Allemagne*, Seconde Partie, XXIII).

It has become familiar ground. "The best and highest that men can acquire they must obtain by a crime" (*The Birth of Tragedy*, Chapter 9). Nietzsche supports his message with three quotations from Goethe—one from his *Prometheus* and two from *Faust.** But in casting the learned doctor as a figure of titanic dimensions, Nietzsche has misread *Faust.* Grasshopperlike, Faust has his ups and downs from the very beginning and talks himself into suicide in the opening scene until saved by Easter bells. The "Forest and

---

*The indefatigable Nietzschean Walter Kaufmann insists on the connection. For the introduction to his translation of *Faust,* Kaufmann writes: "Goethe's opposition to resentful bourgeois morality . . . is quite as deep as Nietzsche's."

Cave" scene interrupts Faust's prospering seduction of Gretchen with a long irresolute monologue on "the austere joy of contemplation" (3239). He is not pleased with his past record.

I stagger from desire to enjoyment
and in its throes I languish for desire.

(3249–50, TR. PETER SALM)

The celebrated lines sound more like a romantic Don Juan than a resolute Prometheus planning glorious exploits. Faust deserves every shaft of Mephistopheles' sarcasm: "The Doctor's in your belly still" (3277); "What a transcendental binge . . . / to inflate one's being to a godlike state" (3282–85).

Careful attention to Goethe's drama suggests that, exposed by Mephistopheles' running mockery of his superhuman pretensions, Faust makes a very distracted Prometheus. He has neither stolen fire nor, like Cain, founded a city. Some years ago, Hans Eichner spotted in Goethe's own writings the maxim that clarifies Faust's true dilemma: "*Der Handelnde ist immer gewissenlos; es hat niemand Gewissen als der Betrachtende.*" "He who acts is always without scruples; only he who contemplates has a conscience." One could restate this moral paradox: Experience is the only route to human knowledge; yet any experience, when reflected upon, incurs guilt. In *Paradise Lost,* Milton has both Adam and Eve find the word *experience* to justify their errant actions. Seen in that light, Faust reenacts the Fall and attains knowledge *(Wissen)* through action, however interrupted and aborted that action may be. The play alternates between action-experience and reflection-conscience.

Faust's problem is that, as a learned doctor, in spite of his attempts to abandon that condition, he can never give himself over completely to resolute action. Thought, reflection, consciousness, scruple—they all interfere with action. At this point, it is almost impossible not to recognize that Faust stands closer to Hamlet than to Prometheus. The solvent power of thinking, of self-awareness, surfaces with *Hamlet* and comes increasingly to haunt literature and philosophy. The conscience-consciousness motif permeates *Faust.* Nietzsche had read both works attentively.

> *Knowledge* [Erkenntnis] *kills action, action requires the veil of illusion—it is this lesson which Hamlet teaches and not the idle wisdom of John-o-Dreams who from too much reflection, from a surplus of possibilities, never arrives at action at all. Not reflection, no!—true knowledge, insight into the terrible truth, preponderate over all motives inciting to action, in Hamlet as well as in the Dionysian man.*
>
> (*THE BIRTH OF TRAGEDY*, Chapter 7, TR. CLIFTON P. FADIMAN)

We cannot just think as we go: we must stop to think.* Not uncertainty about his mother's guilt stops Hamlet; the certainty of it stops him. He knows and cannot cope with the consequences. Where is the connection with Faust? Like Job, Faust knows (and the reader has learned in "The Prologue in Heaven") that he can and will beat the devil and will win final salvation. That knowledge does not liberate; it paralyzes. Three courses are open to Faust, and he declares that he will remain sturdily on course number one.

1. He can live, err, and strive according to our mortal lot: *die Tat.*
2. He can withdraw from life in order to reflect upon his privileged situation.
3. He can choose to do deliberate evil in order to affirm a Satanic or Promethean mode of being.

Having chosen number one in the first "Study" scene, Faust nevertheless shuttles frequently between one and two. Mephistopheles travels at his side, holding out some fairly tame temptations, but Faust never contemplates a course of resolute evil and destruction. He merely bungles things. The damage he does is the unpremed-

*We know that Nietzsche read Emerson, including probably this typically lyric and confusing passage from *The American Scholar* about "the great principle of Undulation in nature": "The mind now thinks, now acts and each fit reproduces the other. . . . Thinking is the function. Living is the functionary. . . . A great soul will be strong to live. . . . This is a total act. Thinking is a partial act." This same motif of thinking versus action tinges every page of Nietzsche's "The Use and Abuse of History" in *Unmodern Observations.*

itated consequence of selfish choices. Has he attained greatness? Some form of tragedy?

In the light of these observations, Spengler's "Faustian man" loses all overtones of Promethean heroism. The motif of striving has become deeply enmeshed in irresoluteness, overweening selfishness, and favorite-son treatment. But Spengler's coinage remains sound. We probably deserve no more heroic a figure on the prow of our ship than flamboyant, bumbling Faust.

Both Faust and Frankenstein deliver us directly into the condition and the problem of excess. Those human beings who leave their mark on others and on history, who stake a claim to some form of greatness, often reach beyond the conventional channels of accomplishment. Riches, power, fame, and sexual adventure represent four extensively overlapping spheres of enterprise through one or another of which many of us can achieve some form of reward. These four primary drives hold out to us a complex area of human activity that under normal circumstances entails no transgression, no forbidden knowledge. But there are those who can experience no lasting satisfaction, who must always reach beyond to a higher tier of drives and rewards, of attractions and repulsions. We can easily cite historical figures to illustrate this Promethean impulse. Alcibiades, Caligula, Cleopatra, Tamerlane, Lorenzo de' Medici, Napoleon—the Athenians coined a word to designate their insatiable greed for the unattainable, for the moon. *Pleonexia* goes beyond common hubris in refusing any limit, any horizon. The four drives of ordinary human accomplishment are abandoned in an aspiration to godhead.

This excess constitutes a problem or a paradox not so much because it afflicts a few unstoppable figures that traverse our lives and our history, but because the rest of us have a hard time not admiring even its most monstrous forms. In the first chapter of *The Civilization of the Renaissance in Italy,* Burckhardt describes the "profound immorality" of Lodovico Sforza, despot of Milan and patron of a brilliant court including Leonardo da Vinci. Burckhardt concludes that the unscrupulous tyrant "almost disarms our moral judgment" by his brilliant contributions to "the state as a work of art." Will Hitler and Stalin have to be added to the above list? Or have we finally learned how and where to draw a line? Let us hope so. But the mythical and barely changing notion of human greatness as passed down from Gilgamesh, say, to Faust and Frankenstein should not set our minds at rest. And it is not difficult to collect statements alerting us to our own proneness to admire forms of *pleonexia.*

> *Only great men can have great faults.*
>
> (LA ROCHEFOUCAULD, *MAXIMS,* number 190)

> *Evil is easy, its forms are infinite; good is almost unique. But there is a kind of evil as difficult to identify as what is called good, and often this particular evil passes for good because of this trait. Indeed one needs an extraordinary greatness of soul to attain it as much as to attain good.*
>
> (PASCAL, *PENSÉES,* Lafuma number 526)

> *He believed he had discovered in Nature . . . something which manifested itself only in contradictions. . . . It contracted time and expanded space. It seemed to be at home in the impossible and to reject, with scorn, the possible. This mode of being I called the Demonic. . . . It appears in its most terrifying form when manifest in a single human being. . . . They are not always the most excellent people . . . but a terrible force comes out of them. . . . From such considerations arise that strange and striking proverb:* Nemo contra deum nisi deus ipse. ["No one can rival God except God himself."]
>
> (GOETHE, *DICHTUNG UND WAHRHEIT,* Part 4, BOOK 20)

These passages do not flinch before the prospect that some form of greatness may lodge in heroes whose conduct has been evil.

Since we seem to be so fascinated by human creatures who aspire to exceed their lot and to attain godhead, how shall we ever reconcile ourselves to a countervailing tradition of heroism in humility and quietism, in finding and in accepting our lot? The line that connects Socrates, Buddha, Jesus, St. Francis, Thoreau, Tolstoy, Gandhi, and Martin Luther King, Jr., has had a hard time restraining human aggressiveness. Consequently, many of us have thrown our support to a third, intermediate set of founding figures who have gradually built our now-besieged institutions of justice, law, and democracy. Since humility has so hard a time restraining hubris, is it possible that our new institutions will begin to afford a new form of greatness in freedom within bounds?

One devoutly hopes so. But Frankenstein and Faust could never resign themselves to remaining in the herd. Their deeply cultivated knowledge of the universe and its secrets filled them not with awe but with *pleonexia*, an overweening resolve to reach beyond limits, particularly limits on knowledge, even at the risk of harming others. In spite of Nietzsche's preachings in favor of the will to power, Faust and Frankenstein cannot be our heroes. Must they, then, be monsters? At least we should be able to recognize that side.

Imagine a literary game in which one is required to assign to famous figures the place they deserve in Dante's three-decker afterlife. Where would Faust go? I find no justification for placing him higher than the adventurous Ulysses in the Eighth Pouch of the Eighth Circle of Hell, the domain of ordinary fraud. *Frankenstein* complicates things for us somewhat by offering us a pair—Dr. Frankenstein as the human monster unwilling to love and nurture his own creature, and the monster himself as (initially) sympathetic hero who did his best to educate himself to become a member of humankind. As he takes care to tell us (653–55; 1112–17), Faust contains the two strains within himself. Their conflict is never fully extruded as dramatic action and remains in the form of words, discussion. Yet for sixty years, Goethe knew he had found the most significant subject of his lifetime, even if he could not do it full justice. The English critic D. J. Enright, having criticized the play's baggy structure, gave a measured verdict. "Impossible though *Faust* is, it is impossible to imagine European culture without it." We all wear *Faust* under our shirt as our most intimate and awkward talisman.

Chapter IV

THE PLEASURES OF ABSTINENCE: MME DE LAFAYETTE AND EMILY DICKINSON

■

The stories that I have discussed so far—Prometheus and Pandora, Psyche and Cupid, Dante's Ulysses, Milton's Adam and Eve, Faust, and Frankenstein—all recount a thrusting aside of limits in a search for knowledge and experience. At the end of the second chapter, I proposed four stages of the downward path to wisdom based on the sequence of events in Milton's version of the (Fortunate) Fall. We may move from ignorance or innocence, to fancy or dream, to experience, to wisdom. Both Dante and Milton placed the word *experience* so carefully in their narratives as to allow us to speak of "forbidden experience" alongside of forbidden knowledge.

But this downward path to wisdom is not the only way in which to live life fully. There is another, less recognized set of stories that approach forbidden knowledge from the other side. They tell not so much of overcoming limits and constraints on experience as of welcoming and taking advantage of them. These tales reveal the rewards of temperance and abstinence over those of indulgence and hedonism. Not prudery and fear impel these stories but, rather, a vital role assigned to the imagination in grasping life and upholding one's identity. Ovid's *Metamorphoses* retells beautifully one ancient parable on the motif of elective, even exultant, withdrawal from experience.

The woodland nymph Syrinx lived in Arcadia and had many suitors. She put them off and escaped from roving satyrs in order to keep her virginity and become like the chaste huntress Diana. One day, Pan, god of fields and forests, saw Syrinx coming down from Mount Lycaeus and desired her. She refused him and fled into the wildest places, until the River Ladon stopped her course. As Pan approached, she implored the other wood nymphs to save her. When Pan arrived at the riverbank, the lovely form of Syrinx was dissolving into tall reeds. He tried to embrace them, but they merely stirred and sighed in the breeze. Pan marveled at the disappearance, and the sweet sound of the reeds charmed him. So he cut some of them and, with wax, bound them together, long and short, to make pipes. He called them syrinx, after the maiden he had lost.

Here is a touching tale. The imaginations of Syrinx in her chastity and Pan in his frustration combine to transform their feelings into music. We do not distort the tale by applying an analytical term to it: *sublimation*. Poets and composers have a great affection for the parable. The former ponder Syrinx's refusal to yield to Pan's advances; the latter usually choose a flute to render the sound of her hollow reeds. For Ovid, metamorphosis into another natural form may represent magic preservation of a sacred state.

Two later stories that reenact this response to forbidden knowledge jar my chronology somewhat. The first belongs to the glorious court of Louis XIV in seventeenth-century France. The other belongs to a small New England town of the mid-nineteenth century. Subtly linked across two centuries, they will modify the significance of forbidden knowledge.

1. Asceticism in *La Princesse de Clèves*

We must beware of standard accounts. In tracing the development of the novel, or perhaps its fall, from idealized romances to particularized realism, literary historians have too often overlooked one of the most significant, enthralling, and psychologically probing novels of the seventeenth century. Mme de La Fayette's *La Princesse de Clèves* was published in 1678, roughly halfway between *Don Quixote* and *Robinson Crusoe.* Such abstract words as *duty, gallantry*, and *esteem*, which characterize its formal style, make the novel sound more like an episode from King Arthur's court than like the investigation of everyday life it really undertakes. We know that it was written by an enthusiastic Sunday novelist and salon hostess linked to the court of Louis XIV. Having borne two children, Mme de La Fayette settled in Paris in 1659. Her husband remained in distant Anjou. At age twenty-five, she considered herself exempt from the inconveniences of love and gallantry. Yet she had an enduring fondness for the great composer of maxims, the Duc de La Rochefoucauld. He probably helped her write the novel. It appeared anonymously with a bookseller's note saying "he" (the author) would reveal himself if the book succeeded with the public. It made a great splash, both before and after publication, but the author clung to anonymity.

Behind its historical facade, *La Princesse de Clèves* explores an eternally contemporary subject: love fright, wariness of deep emotion and of its expression in sexuality. The heroine lives through the essential saga of forbidden knowledge in the domain of romantic love.

Ian Watt in *The Rise of the Novel* has a lame explanation for why he fails to discuss this French novel. He acknowledges its "elegance and concision" and goes on to say: "French fiction from *La Princesse de Clèves* to *Les liaisons dangereuses* stands outside the main tradition of the novel . . . we feel it is too stylish to be authentic." It is precisely because her stylishness accomplished an authentic portrayal of the women and men of her milieu that we should be impatient with Watt's summary dismissal.

One further reason why a history of the novel is incomplete without Madame de La Fayette's masterpiece lies in its challenging action. Every detail and digression in *La Princesse de Clèves* helps explain how one woman's aching indecisiveness about her life moves gradually toward the resoluteness she finally achieves. Carefully trained and educated before being presented at court at age sixteen, the future Princesse de Clèves marries an excellent man who loves her very much and wins her esteem, not her love. Later she meets the Duc de Nemours, the most gifted and attractive nobleman in the King's entourage. Though they barely exchange a word during the balls, jousts, and salon gatherings of life at court, these two paragons fall in love "by fate." In a scene that has become famous, the Princess brings herself to confess her love to her husband without naming its object. One implausibility is matched by another: The Duc de Nemours himself is eavesdropping outside the window. Great tension builds up on both sides of the marriage. Nevertheless, when the Prince de Clèves finds out from other sources that his rival is the Duc de Nemours, the discovery leads to another astonishing exchange, or, rather, to an unforgettable silence. The Prince de Clèves is speaking to his wife while they are alone in her room.

> *"Of all men the Duc de Nemours is the one I was most afraid of, and I see your danger. You must control yourself for your own sake and, if possible, for love of me—I don't ask it as a husband, merely as a man whose happiness depends on you and who loves you even more tenderly and passionately than you love that other man."*
>
> *As he spoke, the Prince de Clèves broke down and could hardly finish what he was saying. His wife was penetrated to the heart, and bursting into tears she embraced him with such tender sorrow that his mood changed a little. They stayed like this a while and separated without having spoken again; indeed they had no more strength for words.*
>
> (132–33, TR. NANCY MITFORD, MODIFIED)

Only a confident author knows when to renounce the lifeblood of narrative: words. Here, that authorial renunciation relates closely to the action unfolding around the stricken Princess.

False information implying his wife's unfaithfulness causes the Prince de Clèves to fall ill. Before he dies, she almost convinces him of her virtue. In due course, the Duc de Nemours presses his suit again. Nothing now stands in the way of the Princesse de Clèves accepting the pleasures of reciprocated passionate love under favorable conditions and with everyone's approval, even the King's—nothing, that is, except her remorse over having contributed to her husband's distress and death, and her sense of duty. Her "scruples" go very deep.

The Duc de Nemours arranges a surprise meeting with the Princesse de Clèves alone. Summoning all her courage, she acknowledges that she returns his love but that she cannot face the possibility of seeing his sentiments for her diminish with time. At the climax, she hides nothing and refers loyally to her husband, who has died of love for her. She is pleading for something as rare in life as in fiction: integrity of feeling, a blend of passion and lucidity. It controls the smoldering words she addresses to the Duc de Nemours during this final interview.

> *There was perhaps one man and one man only capable of being in love with his wife, and that was M. de Clèves. It was my bad luck that this brought me no happiness—possibly this passion of his would not have continued so strong if I had requited it, but I cannot use that means for keeping yours. Then I have an idea that it was the obstacles which kept you so true to me.*
>
> (192)

Her controlled impetuousness hits every nail on the head. She holds firm against the Duc de Nemours's impassioned pleading for their marriage and maintains that by renunciation, her feelings for him will not die. As in the story of Héloïse's violently enforced separation from Abelard, this elected separation leads not to the displacement of feelings we call sublimation but to an intensification of response related to fanaticism and idolatry.

That night, the Princesse de Clèves examines her situation. Some of the analytical language in this passage has been used earlier to describe how someone falls in love, especially the word *étonnement,* "astonishment." It means a sudden and wrenching self-beholding. We are almost at the end of the novel.

> *There was no peace for Madame de Clèves that night. After all, this had been the first time she had left her self-imposed retreat as well as the first time she had ever allowed anybody to make a declaration of love to her; added to which, she herself had now admitted to being in love. She did not recognize herself anymore. She was amazed to*

> *think what she had done, and, not knowing whether to be glad or sorry, her mind was filled with a passionate restlessness. She went over the reasons which duty seemed to put between her and her happiness, and found, to her sorrow, that they were very powerful; she wished now that she had not described them so lucidly to M. de Nemours.*
>
> (197)

It is not difficult to see why this has been called the first psychological novel, a category usually reserved for the following century. This kind of introspective analysis takes the place of the classic stage scene with a confidant and anticipates the probings of interior monologue. The Princesse de Clèves is amazed at herself, even irritated with herself, on two counts. She has told the truth to the very person from whom decorum requires she withhold it. Equally remarkable, she has acknowledged most of the truth to herself. Her feeling of "astonishment" represents the shock of self-consciousness. That state does not free her to follow her inclinations; it obliges her to recognize how complex her inclinations have become. In these concluding pages, she finds a higher selfishness (to remain a widow rather than to risk the pangs of jealousy in marrying the Duc de Nemours) that coincides with a higher duty (to shun the man implicated in the death of her husband). To realize her love would, she fears, destroy it. She will preserve it by suspending it in the amber of her past. The novel ends undramatically with a long journey followed by a longer illness and partial retreat to a nunnery. In calm, formal sentences, we are informed that she finds peace of mind before she dies.

Soon after publication in 1678, *La Princesse de Clèves* was engulfed in two vigorous controversies. One concerned its genre. The *roman*, or "romance," usually dealt with high chivalric or pastoral adventures described in an inflated style and often included implausible and supernatural episodes of shipwreck and families miraculously reunited. The *nouvelle* favored simpler, shorter narratives that developed less extravagant codes of conduct. This anonymous story presented the seemingly fantastic action and personages of a *roman* in the down-to-earth settings and style of a *nouvelle*. The controversy about the book's *vraisemblance* ("plausibility," "believability," "verisimilitude") covered much the same ground and focused on a few celebrated scenes, most of all on the scene of the avowal. Would or should a well-behaved wife ever confide to her husband that she had fallen in love with another man? Contemporary maxims could be quoted on both sides: A wife should never alarm her husband; a wife should tell her husband everything. To this day, critics do not agree to what extent Mme de La Fayette's episodes overtax our credulity and weaken the novel.

Behind its reliance on psychological and narrative conventions still far removed from realism, I find *La Princesse de Clèves* revealing and convincing as a kind of pedagogical novel. The resolute character of the Princess and the recognition of her virtues by two exceptional men throw into relief the importance of her education. Innocence must be prepared for the trials and corruptions of life at court through the telling of appropriate stories fortified by maxims and rules. Accordingly, the book is full of narrative digressions, which are really cautionary tales about the depredations of love. How much should an aristocratic young girl be told? Does knowledge about the temptations of the world temper the passions? Or does it arouse them? Mme de La Fayette believes in full disclosure. Therefore, with all its stylization, the novel tells a great deal about life at the French court in the seventeenth century.

The Princesse de Clèves is not a saint. Her asceticism appeals more to psychology than to religion. Human, not spiritual, motives impel her to renounce what she most passionately desires. She will not choose pleasure in the short run because, if she does so, she foresees suffering and despair in the long run. Her difficult yet resolute decision springs as much from an instinct for survival as from strong moral feelings. This residual, self-protective selfishness esteems the mysteries of love more than it rejects them.

We can appreciate the singularity of this attitude by comparing Mme de La Fayette's novel to two celebrated epistolary novels of the eighteenth century, Rousseau's *La nouvelle Héloïse* (1761, a runaway best-seller) and Choderlos de Laclos' *Les liaisons dangereuses* (1782). In a hundred years, France changed from a conformist society that supported an absolute monarchy enacting its daily rituals on the stage of Versailles to a decaying aristocracy opposed by a strong bourgeoisie and by an articulate band of freethinking *libertins* and *philosophes* who criticized religious and political traditions in the name of reason and nature. One of the *libertins,* Rousseau wrote his novel about a passionate yet submissive woman torn, like the Princesse de Clèves, between attachments to two men.

In Book XI of his *Confessions,* Rousseau brags about his enormously successful *Julie,* or *La nouvelle Héloïse*: "Without fear I place its Fourth Part alongside *La Princesse de Clèves.*" Rousseau's two-volume saga explores social and emotional terrain that he considered an extension of Mme de La Fayette's confined universe. Like Abelard and the original Héloïse, like Paolo and Francesca, Julie and her tutor, Saint-Preux, fall in love and briefly become lovers. Her father has promised her hand to a worthy friend, Wolmar. After her mother's death, caused by her discovery of Julie's lapse, Julie feels she must obey her father. Saint-Preux proposes secret, virtuous adultery. Julie undergoes a "revolution" and finds in honor the motive of virtue. "Yes, my good and worthy friend," she writes to Saint-Preux, "in order to love each other forever we must renounce each other. Let us forget all the rest; be the lover of my soul. So tender an idea is a consolation for everything else." Tens of thousands of eyes across Europe wept over the passage. The story is only half-told.

Happy among her children, Julie confesses everything to her understanding husband, Wolmar. Saint-Preux comes to live—chastely—near their estate, where total frankness creates an open society, a model farm, "a house of glass," and an apparently ideal ménage à trois. A few years later, Julie's last letter to Saint-Preux, written on her deathbed, after a long illness, carries the situation one step further. She still loves him passionately; her temporary "cure" saved both her virtue and their love. "The virtue that kept us separate on Earth will unite us in the eternal life."

Unlike the Princesse de Clèves Julie finds a way to renounce her cake and to have it, too. In her transparent household, duty and honor do not have to suppress all forms of exaltation in forbidden love. Healthy sublimation? Rousseau hopes so. Yet for all her gushing feelings, Julie relies on a half-repressed hypocrisy to sustain in her marriage a fantasy adultery. The Princesse de Clèves firmly avoids such sentimental complications by retiring to a convent.

Les liaisons dangereuses, published on the lip of the Revolution, depicts a milieu not of sentimentality but of extreme cynicism. A trained soldier, later a general under Napoleon, Laclos wrote about the Machiavellian sexual connivings of two depraved aristocrats without a court who seek to take revenge on former lovers and to besmirch any innocence or virtue they encounter. These freelance predators, Valmont and Madame de Merteuil, admit no duties or scruples to restrain their desires. Love is reduced to a series of competitive power plays described in their letters with chilling cruelty and vanity. As with Molière's *Dom Juan*, all satisfaction arises from conquest—sexual, intellectual, moral. Every conquest leads to new levels of jealousy and envy, which undermine the very possibility of attachment. The contemptuous and sometimes bantering style of the letters makes it difficult to decide whether Laclos is condoning or condemning the exploits of his two dedicated *libertins*. The Marquis de Sade was reaching maturity in this society of systematic depravity.

La nouvelle Héloïse and *Les liaisons dangereuses* describe the deflection of love in the eighteenth-century novel into extremes of sentimentality and cynicism. A century earlier, both the passion and the calculation portrayed in *La Princesse de Clèves* have more intensity than those elements as represented in the later novels. Neither Rousseau nor Laclos could occupy the psychological space opened up by Mme de La Fayette in a much shorter work than either of theirs.

Mme de La Fayette's portrayal of a woman's fear of compromising her love by consummating it cannot be dismissed as a period piece, an old-fashioned story, an aberration. The Princesse de Clèves does not, as some students have suggested to me, lose her mind after her husband's death. Nor do I find evidence that her marriage has remained unconsummated or that she is frigid. She feels, rather, the impulse to withdraw from intimate encounter with a person toward whom she is attracted by passionate love. The impulse to withdraw blends psychological and moral scruples into what I have referred to as higher selfishness and into a story of undeniable tragic force. That story does not deny love, but internalizes it and cherishes it—while suffocating it, some would say.

On the downward path to wisdom, the Princesse de Clèves resolves to reach wisdom without the stage of experience, relying on her imagination to close the gap. She seems to grasp the immensity of the challenge. As one might expect, there is not a huge number of literary works that explore this austere moral lucidity—or blindness. We are drawn more to the lives of sinners than of saints. Yet stories about renunciation of love have held a secure place across the centuries and provide the full setting for Mme de La Fayette's novel.

In the sequence of speeches on love that make up Plato's *Symposium,* or *Banquet,* Socrates does not speak last. That position is reserved for Alcibiades' half-drunken account of how Socrates rejected his amorous advances—that is, gently and firmly refused the proffered love of a strikingly handsome warrior, still youthful and already celebrated. Socrates, beautiful in his own right and not unmoved, honors love by knowing when to decline its physical expression. The *Symposium* "opens up" Socrates like a nested doll, to reveal a remarkable moral agent whom the "sacred frenzy" of philosophy leads not to debauchery but to abstemiousness.

Likewise, all George Eliot's novels concern renunciation in some form. In the most melodramatic of them, *The Mill on the Floss* (1860), a young woman as beautiful and as ardent as the Princesse de Clèves turns down two men in favor of deeper ties represented by her upright brother. "I cannot take a good for myself that has been wrung out of their misery" (Book VI, Chapter 14)—that is, out of hurt inflicted on friends and family. But in their immense lucidity both Maggie, the heroine, and Eliot, her creator, know that a major decision like renunciation will not solve everything. "The great problem of the shifting relation between passion and duty is clear to no man who is capable of apprehending it" (Book VII, Chapter 2). This profoundly paradoxical sentence deserves long consideration and leads us close to forbidden knowledge in its most intimate form. To call this complex notion moral agnosticism improves not a whit on Eliot's carefully turned sentence or on the vital novel that contains it. The sentence in context also affirms that no moral abstraction or maxim will provide a "master key" to any such dilemma. One must know the full story in all its human circumstances—as Eliot here provides.*

Compared to the five hundred full-blooded pages of *The Mill on*

**The Mill on the Floss*, like *La Princesse de Clèves*, can be seen as a foil to the two great modern novels about experience not rejected but seized: *Madame Bovary* and *Anna Karenina*. All four books introduce reading and stories as essential sources for the heroine's response to romantic love. Emma and Anna succumb very young to the allurements of sentimental novels. Not so the other two heroines. Starting well

the Floss, the decadent-symbolist drama *Axel* (1890) reads like a cartoon. Yet the half-forgotten work has genuine significance in the present context of renunciation. Edmund Wilson's *Axel's Castle* (1931) takes its title and its theme from this play by Villiers de l'Isle-Adam.* Count Axel of Auersburg lives in austere isolation in a Gothic-Wagnerian castle in the Black Forest, a castle beneath which is concealed a vast treasure. When Sara, a mysterious young intruder of noble blood, discovers the treasure, Count Axel catches her in the act. They fight with pistols and daggers, survive with insignificant wounds, and of course fall passionately in love. They now have everything, including each other. The world lies before them. The last scene in the crypt of the castle carries the title "The Supreme Option." In the crucial passages, it is difficult not to hear a parodic echo of *La Princesse de Clèves*, with the sexes reversed.

> SARA: *Axel!* [He is pensive.] *Axel, are you forgetting me already? The world is out there. Let's go live!*
>
> AXEL: *No. Our existence is already fulfilled. Our cup runneth over. All the realities, what will they be tomorrow compared to the mirages we have just lived?*

His speech goes on a long time and remains deadly serious. Axel's most famous line is in no way meant as a joke. "Live? Our servants can do that for us." Sara and Axel poison themselves without consummating their passion, thus affirming the primacy of imagination over reality.†

---

before her marriage, the Princesse de Clèves hears a series of cautionary tales about the perils of love among the nobles at court. Maggie gives back the romantic novel *Corrine* unfinished to Philip, for it is the religious meditations of Thomas à Kempis that arouse in her "a strange thrill of awe" (Book IV, Chapter 3).

*Wilson's chapters offer the earliest and best examination of what we now, for lack of a better term, call "modernism." He called it Symbolism.

†One recent enactment of the renunciation story is surely destined for a literary or stage version. For several years, Suzanne Farrell was the favorite in George Balanchine's New York City Ballet—the favorite dancer and the favorite woman. She refused to become his lover or wife and threw herself into dancing. Her description of the long encounter rings true and brings on stage a modern-day princess with somewhere to go other than a convent or her grave. "Our unique relationship had proved itself . . . often to both of us, and it might not have withstood consum-

Wilson chooses this stilted yet impressive drama to represent a major aspect of the symbolist attitude: *withdrawal from life into thought and language.* Axel's refusal to run the risk of living corresponds to a profound current in the symbolist attitude toward language. Musicality, delicacy, deliberate obscurity, *la chanson grise*—all these elements of symbolist poetry represent an extreme point in the history of Western literature. The essential poet of this amorphous movement, Mallarmé, set down the most succinct statement of the symbolist approach to language: "*To name* a thing is to destroy three-quarters of the poem's enjoyment, which consists in getting at something little by little, in gradually divining it. The ideal is *suggestion*" ("Sur l'évolution littéraire," 1869). This pronouncement deserves thought beyond the confines of symbolist doctrine. Mallarmé's two sentences imply that there are feelings and states of mind so delicate as to be best approached indirectly, by mere hints, by evocation in sound and sense. If I use a word so explicit, so obvious as, for example, *embarrassment* or *anger*, I reduce a complex psychological state to a stereotype, to a convention we think we share, to a caricature of itself. Another poet, Paul Valéry, drew the full conclusion: "To see a thing truly is to forget its name." The most exciting enterprise of language is to avoid using language according to its conventional forms. Don't make anything too clear. The imagination needs a milieu of mystery to work in. Flaubert was talking about the same kind of literary purity when he refused in outrage to allow *Madame Bovary* to be illustrated. That would be worse than naming names! For all her reliance on abstract psychological words such as *esteem* and *duty*, I believe that Mme de La Fayette displays in *La Princesse de Clèves* a sense of this withdrawal from naming when she uses the word *étonnement* to des-

---

mation. The physical side of love is of paramount importance to many people, but to us it wasn't. Our interaction was physical, but its expression was dance." The critic Mindy Aloff, who cites the preceding sentences, was astute enough to call Farrell "a heroine . . . of her own imagination."

One wonders what the connection may be between the Balanchine-Farrell story and the case of Edward VIII, who abdicated the throne of England in 1936 in order to marry a commoner and divorcée. Claude Sautet's music-filled film *Un coeur en hiver* (1992) tells the story of a woman's love refused by a man who half-believes that such feelings do not exist. Everyone and everything else in the film, including Ravel's sensuous music, belies his attempt at emotional isolationism.

ignate what happens to her heroine when subjected to shock or passion. Like the symbolists, Mme de La Fayette had a strategy: At key points, don't say it; suggest it. Her stylishness (if that is the right word) consists essentially in a rare subtlety of expression, an aesthetics of discretion.*

Someone told me as a child how to see a star at night: Don't look directly at it; look slightly to one side of it. It was years before I learned about the physiology of rods and cones on the retina. This indirect approach to the subtleties and complexities of the world lies at the heart of symbolism as described by Mallarmé. I believe that we can go at least one step further. Both in the tumultuous self-denials of *La Princesse de Clèves* and of *Axel,* and in the emotional and stylistic reticence of symbolism, one can discern a state of mind in which asceticism or self-denial approaches close to aestheticism, the cultivation of art and beauty. For both asceticism and aestheticism entail an activity of the imagination that is the contrary of closing one's mind.

2. Aestheticism in Emily Dickinson

With my kitful of stories, I have been pursuing forbidden knowledge in this chapter as it takes the shape not of bold curiosity but of self-restraint and withdrawal. In that context, eight lines of a single poem by Emily Dickinson, because they describe the rewards of renunciation, bear comparison with Mme de La Fayette's two-hundred-page novel. Few pairs of works provide so striking a contrast between the dimensions and dynamics of short lyric and extended narrative. We must first approach Dickinson's poem unhurriedly and without disturbance, as we would approach a brook trout lurking in a pool.

In 1862, at age thirty-two, Dickinson learned that the celebrated Philadelphia pastor Charles Wadsworth had been called to a new church in San Francisco. There is strong evidence that seven years

*The impulse toward indirection and suggestion has also contributed, I believe, to developments in art from Impressionism and Cubism to abstraction. (See my "Claude Monet: Approaching the Abyss," in *The Innocent Eye.*)

earlier, when she had heard him preach and had met him in Philadelphia, Dickinson fell deeply in love with the eloquent clergyman. They corresponded. She may have addressed and even sent to Wadsworth the three astonishing "master" letters of which drafts were found among her papers. He called on her in Amherst in 1860 while visiting another friend in the vicinity. Apparently, the happily married clergyman sixteen years her senior did not reciprocate her intense feelings.

After his departure by sea for California via Cape Horn, Dickinson assumed the life of a recluse in a white gown, entered the most productive period of her poetic career (a poem a day for over a year), and took the uncharacteristic step of sending out a few of her poems to a stranger. She chose as her mentor Thomas Wentworth Higginson, a young Unitarian clergyman and abolitionist agitator who had just contributed to the *Atlantic Monthly* an article of encouragement to young American writers.

"Are you too deeply occupied to say if my Verse is alive?" So ran her opening sentence to Higginson. Like a valentine, this first letter in tiny birdlike writing, with four poems enclosed, carried no signature. She had printed her name faintly in pencil on a card sealed inside a separate envelope also enclosed. Higginson, who had the force of character to take command of the first Negro regiment in the Union army a few months later, accepted the mysterious woman's challenge and ventured to make a few criticisms along with some inquiries of his own. Dickinson's second letter to him blends coquettishness, literary unorthodoxy, wicked wit, and sheer hallucination into a document so subtle and so blunt that it must be read complete. Every sentence is drawn up out of a deep cistern of accumulated experience.

25 April 1862

Mr. Higginson,

Your kindness claimed earlier gratitude—but I was ill—and write today, from my pillow.

Thank you for the surgery—it was not so painful as I supposed.[1]

1. Higginson's comments took issue mostly with her unconventional orthography and word usage.

I bring you others—as you ask—though they might not differ—

While my thought is undressed—I can make the distinction, but when I put them in the Gown—they look alike and numb.[2]

You asked how old I was? I made no verse—but one or two—until this winter—sir—.[3]

I had a terror—since September—I could tell to none[4]*—and so I sing, as the Boy does by the Burying Ground—because I am afraid—You inquire my Books—For poets—I have Keats—and Mr and Mrs Browning. For prose—Mr Ruskin—Sir Thomas Browne—and the Revelations.*[5] *I went to school—but in your manner of the phrase—had no education.*[6] *When a little Girl, I had a friend who taught me Immortality—but venturing too near, himself, he never returned— Soon after, my Tutor died—and for several years my Lexicon—was my only companion.*[7]*—Then I found one more—but he was not contented I be his scholar—so he left the Land.*

You ask of my Companions Hills—Sir—and the Sundown—and a Dog—large as myself, that my Father bought me—They are better than Beings—because they know—but do not tell—and the noise in the Pool, at Noon—excels my Piano.[8] *I have a Brother and a Sister—My Mother does not care for thought—and Father, too busy with his Briefs—to notice what we do—He buys me many*

---

2. Dickinson frequently used the words *thought* and *mind* as synonyms for "my poems." The coy metaphor of attire used here probably compares the sketchy, endlessly amended rough sheets on which she composed her poems with the fair copies made for Higginson.

3. A double evasion of Higginson's question. Dickinson had been writing poems for at least three years and had produced over two hundred of them, including several now celebrated—for example, "I never lost as much but twice" and "I taste a liquor never brewed."

4. Probably refers to Wadsworth's move to California. Dickinson had no reason to hide her fears about her eyesight. They afflicted her at about the same period.

5. Principal omissions: Shakespeare, Emerson, Thoreau.

6. Dickinson attended Amherst Academy for six years and Mount Holyoke Female Seminary for one. She was particularly known for her wit and funny stories.

7. "Friend" and "tutor" may refer to two young men who encouraged her literary interests and died young: Leonard Humphrey, principal of Amherst Academy, and Benjamin Newton, a law student in her father's office. Samuel Bowles and Charles Wadsworth cannot be excluded from these oblique allusions, especially "one more" in the following sentence.

8. The surprising words appended after the "tell"—testifying to both literal and hallucinated perceptions—bear comparison with images in the "Alchimie du Verbe" section of Rimbaud's *A Season in Hell* and with "le déréglement de tous les sens" in his 1871 *Lettre du voyant.*

books—but begs me not to read them—because he fears they joggle the Mind. They are religious—except me—and address an Eclipse, every morning—whom they call their "Father."[9] *But I fear my story fatigues you—I would like to learn—Could you tell me how to grow—or is it unconveyed—like Melody—or Witchcraft?*

You speak of Mr Whitman—I never read his Book—but was told that he was disgraceful—

I read Miss Prescott's "Circumstance," but it followed me, in the Dark—so I avoided her—[10]

Two Editors of Journals came to my Father's House, this winter—and asked me for my Mind—and when I asked them "Why," they said I was penurious—and they, would use it for the World—

I could not weigh myself—Myself—

My size felt small—to me[11]*—I read your Chapters in the Atlantic—and experienced honor for you—I was sure you would not reject a confiding question—*

Is this—Sir—what you asked me to tell you?

Your friend,

E—Dickinson.

Though he admired her poetry and responded to her letters, Higginson's comments had little effect on her work. Nevertheless, Dickinson later wrote him that "you saved my life" and urged him to visit her. One of the poems she sent to him with her third letter contains the lines "Renunciation—is the Choosing / Against itself—." She never sent him a later poem on the same theme that simultaneously opens and closes the curtain on her inner life. It is these eight lines of poem number 421 in the Johnson edition that place Dickinson's poetic persona alongside the fictional seventeenth-century Princesse de Clèves. I urge the reader to read the poem several times, preferably aloud, before going on.

---

9. One could extrapolate from this paragraph the essence of Dickinson's family relations, her courageous intellectual life, and the complex evolution of her religious beliefs. In her sometimes mocking skepticism, she never gave up faith in immortality.

10. Harriet Prescott Spofford contributed "Circumstance" to *Atlantic Monthly* for May 1860.

11. These two sentences of eight and six syllables respectively have all the earmarks of the opening of a formal poem in pure Dickinsonian diction.

A Charm invests a face
Imperfectly beheld—
The Lady dare not lift her Veil
For fear it be dispelled—

But peers beyond her mesh—
And wishes—and denies—
Lest Interview—annul a want
That Image—satisfies—

One of Dickinson's simpler poems, "A Charm" displays glints and recesses that will enlighten us about the nature of poetry while conveying an intricately constructed meaning and an implied narrative. The desire shared by many human beings to assemble meaningful sounds into a gemlike utterance appears to result from maintaining into adulthood two childhood stages of language familiar to us all: babbling, or lallation, and the punning riddle. At some time after six months, a child begins to hear and say sounds in repetitive patterns that prepare the way for rudimentary nursery rhymes like "Hickory, Dickory Dock" and "Fee, Fi, Fo, Fum." From about the age of six years on, the child revels in dumb riddles based on puns, on words that reveal the hidden interconnections and short circuits of our language. What's good for water on the brain? A tap on the head. When is a door not a door? Repetition (babbling) and transformation (punning) offer a whole universe to play with. Then the play becomes very serious. When these two instinctual responses to language combine and develop, they provide the territory of poetry. Dickinson's enormously sophisticated and condensed composition yields much of its significance in these two categories.*

By the time the English language was carried to Puritan New England, infant babbling had been regularized into a variety of traditional forms from nursery rhymes to the sonnet. In "A Charm," Dickinson uses a pattern familiar to her from church services she

*My contrast between the processes of repetition and transformation parallels that of Hume for association of ideas (contiguity and similarity), of Freud for dream work (condensation and displacement), and of Jakobson for poetry (metonymy and metaphor).

attended weekly during her formative years. In order to facilitate their musical setting, hymn verses were classified into a limited number of standard schemes according to syllable count. *Common meter* alternates lines of eight and six syllables, as in Tallis' Ordinal:

> *The great Creator of the worlds,*
> *The sov'reign God of heaven . . .*

Long meter has four lines of eight syllables each as in Tallis' Canon:

> *All praise to thee, my God, this night,*
> *For all the blessings of the light.*
> *Keep me, O keep me, King of Kings,*
> *Beneath thine own Almighty wings.*

Stress and feet concern us far less in these stanzas than what Milton and Pope called "numbers"—strict syllable counting. Working still in a developed form of lallation related to the jingle, Dickinson chose *short meter* as her model for "A Charm": 6,6,8,6. A few readers will find a corresponding hymn tune right away ("Franconia"):

> *The ancient law departs,*
> *And all its terrors cease.*
> *For Jesus makes with faithful hearts*
> *A covenant of peace.*

Such a stanza draws a deep breath in the third line and then settles back to the basic beat. The pattern can go on and on, as in a ballad.

Compared to the hide-and-seek syntax used in many of Dickinson's poems, the four clearly articulated clauses in this poem present little difficulty. The sense of a riddle needing solution hovers over the poem as a whole. Almost every word, as we shall soon see, is a real or incipient pun. The only moment of syntactical uncertainty comes in the fourth line with the unremarkable *it*. Normal parsing connects it to *Veil*; then the reader must revise the rules and look all the way back to *Charm* for an adequate antecedent.

The essential mystery of the poem circulates through such half-declared queries as: Whose face? What is the situation? What tone of voice? How marked a shift between the two stanzas? I believe word-by-word commentary will address these questions more effectively and concretely than overall interpretation at the outset.

Charm: The word refers to a range of forces, from physical attractiveness to magic and witchcraft. Dickinson's letter to Higginson suggests how deeply she responds to all those forces. *Charm* inspires *want* (7) and thus projects its presence forward through the whole poem. Capitalization helps to reveal the word our ear and eye tell us lurks behind Charm: *harm*. In that embedded opposition, *charm / harm*, the whole poem lies latent. Later oppositions recapitulate this one, which implies a presence both attractive and forbidden.

invests: The word means "to put on like a vestment or garment"; "to instill"; "to install." Also, secondarily in 1862, the meaning is "to employ money for interest or profit." Thus we encounter a sacred meaning shaded by a profane one, and even further tainted by a hovering rhyme and near homonym: *infest*. See above. The *harm* motif is reinforced by this further echo.

a: Two indefinite articles in the first line serve to distance and generalize the implied scene. That effect will change.

face: The word is not capitalized. It suggests outward appearance only, as compared to a word like *physiognomy*, which expresses inner character. Whose face? We don't know on first reading. On reaching *Lady*, we provisionally attribute it to her. Further readings reveal that the first sentence is ambivalent: *Charm* resides either in her own veiled face or in the other face she observes through the interference of her veil, or, more probably, in both faces.

Imperfectly: What we see too clearly loses its charm. The poem turns on a valued impediment to full perception. Dickinson glosses this crucial adverb in another poem (number 1071), which opens: "Perception of an object costs / Precise the object's loss—."

beheld: This word means "to see," "to apprehend," "to possess (as in holding something)." The last sense provides the kick for the impending rhyme.

The: The definite article is used now to particularize the situation after two indefinite articles.

Lady: Capitalized as in allegory, the noun propels us both toward worldliness of elevated social position and toward holiness, virginity, as in Our Lady.

dare: The word introduces the note of risk and fear that is extended in the following line and prepares us for the difficult decisions of the second stanza.

Veil: The capitalized word combines two conflicting associations of concealment: purity and retreat to the cloister in "taking the veil"; and latent coquettishness, flirtation, in hiding behind a veil in order to look out more freely. Dickinson was always fond of the word. In 1853, aged twenty-three, she wrote her friend Susan Gilbert, "I find I need more vail [sic]."

fear: See *dare* above.

it: The word grammatically refers to *Veil*, which meaning is weakly maintained. When combined with its verb, the antecedent gravitates strongly backward to *Charm* and then moves outward to embrace the whole poem. The second stanza informs us—like the close of *La Princesse de Clèves*—how to dispell the *fear* of losing the precious *it*.

dispelled: An opposite and equal force to *beheld*. The powerful sense of driving away, of dispersing, should not be diluted by any fantasy formation isolating the word *spell*, even though it reasserts the magical side of charm.

But: This is the obvious pivot on which both sound and sense turn. Why not *And*? Logically constructed, the first stanza calls for the behavior described in the second. The transitional word could well be *Accordingly* . . . in a prose version. Still, we understand that Dickinson wants to insist on a contrast, a shift of direction within the overall unity to suggest that the actions of the second stanza require courage. The Lady does not dare to lift her Veil *but* she dares to look out boldly and shrewdly upon the scene. *But* instead of the equally correct *And* frames and stages the melodrama of the second stanza by animating both the cloistered nun and the (repressed) flirt.

peers: The word proffers a beautiful non-Latinate monosyllable. It rhymes with *fears*. To peer suggests having to expose oneself just a little.

beyond: More than *through* or *around*, *beyond* implies a limit, a line drawn in the moral and psychological landscape.

mesh: The word designates the interstices of a net or veil, the way through it, while connoting entanglement, imprisonment. It is probably related etymologically through Old Norse to *mask*.

wishes: The simplest of words here expresses the full range of physical and spiritual yearning, and is just as strong as the current cant word: *desire*.

denies: Compared to *restrain* or *hold back*, this word implies a very categorical action, with biblical associations from Peter's triple denial of Jesus. It prepares us for the opposition of sense in the rhyme to come.

Lest: The conjunction repeats almost exactly the meaning of *For fear* (4) and leads this time to a succinct explanation of the *Lady*'s behavior.

Interview: Here, the word means to meet face-to-face without any veil or impediments. Highly visual, personal, and physical, it moves beyond the glimpsing of French *entrevoir* to full exposure.

That: The relative pronoun becomes confusing and misleading if read as a demonstrative adjective.

Image: The word evokes a mental representation or simulacrum provided by the faculty of imagination. As the opposite of *Interview*, *Image* implies that the *Lady* chooses both absence and abstinence.

satisfies: This is a defiant, even triumphant affirmation with which to close a poem about resisting temptation.

NO PARAPHRASE OF a poem will suffice. On the other hand, every attentive paraphrase contributes to our understanding. Looking through her veil, a woman feels deeply drawn by the almost-magical beauty of another person, for whom she, thus concealed, may herself exert a powerful charm. She decides to place her faith in the picture she can represent in her mind rather than to seek fuller or more intimate knowledge of the other person.

Only three words in "A Charm" exceed two syllables. Dickinson condenses a potentially fulsome story by squeezing it between two pairs of opposite terms, which are also her rhymes: *beheld / dispelled* and *denies / satisfies*. Thus her versification does not merely decorate; it expresses in sounds the dynamics of the implied action. She also reinforces her impeccable rhymes by the capitalized contrast in the last lines: *Interview / Image*. Exploiting the ambiguities of the key word, *Veil,* Dickinson has given to a traditional subject a form so concise as to look like an epitaph. An aphoristic version of the theme might read: The imagined surpasses the real. The historic anti-Enlightenment outburst of feeling in the nineteenth century known as Romanticism clutched to itself this deep—and sometimes desperate—faith in the products of the imagination. For a century and a half, any bright student of English literature has been able to recite the locus classicus of the theme:

> *Heard melodies are sweet, but those unheard*
> *Are sweeter . . .*

Keats sings to persuade himself and us that the "marble men and maidens overwrought" on the Grecian urn have overcome mortality and contingency. "Cold Pastoral!" must be read half ironically. Since the bold lover will never finally kiss the painted maiden, "ever wilt thou love and she be fair." In the world of Keats' painted figures, ecstasy and beauty will never be dispelled. Likewise, Dickinson's protagonist, reminding us of the Princesse de Clèves, chooses the ideal over the real. Her lines offer far more than an autobiographical situation tempting us to identify the actors. Dickinson has transcended the personal without having to renounce the vividness of a specific scene. Nor would we do justice to "A Charm" by insisting that it primarily records or celebrates poetic creation, the act of composing this poem. The central movement concerns not verbal creation or expression, but the dynamics of the *Veil*—taking it, taking advantage of it, (not) lifting it, being both caught and freed by it.

The strongest challenge of this tiny work occurs when one proposes to read it aloud—the true test of reader, poem, and listener. What tone of voice will do it justice? "A Pen has so many inflections and a Voice but one." This sobering sentence from a letter by Dickinson to Higginson in 1876 is not a plea for writing and silent reading, but a plea to find out the true inflection of a real voice—in this letter, an appeal to *Interview* over *Image.* A reading aloud does not and cannot exhaust the tonal possibilities; it enacts some of them. How then shall one read these lines aloud? Coquettishly? Fearfully? Sentimentally and then—pivoting on the *But*—resolutely? Reflectively, almost neutrally, letting the individual words elicit their multiple and countervailing effects? The last, I believe, works best, with just a trace of archness to hint at the other possible moods: A slow reading, responding to every dash and capitalization.

Many years after "A Charm," Dickinson found the sumptuous phrase "The Banquet of Abstemiousness" (number 1430) for her theme. We should enjoy that banquet "lest the Actual— / Should disenthrall thy soul—." Both poems counsel not against desire but against yielding to desire without fully consulting the soul's scruples. A reasonable asceticism contributes to the aesthetic delectation of life.

And now, if I have done my work adequately, it should be possible to quote two further stanzas without commentary.

Heaven—is what I cannot reach—
The apple on the tree
Provided it do hopeless hang—
That—Heaven is to me—

The color on the cruising cloud—
The interdicted ground
Behind the hill—the house behind—
There—paradise is found—

(1377)

3. An Epicurean at "The Banquet of Abstemiousness"

In her letter soliciting help from Higginson as well as in her *Veil* poem, Dickinson gently yet firmly resists full revelation, full knowledge. Don't hope to learn my exact age, my entire appearance, my inner soul. Let your imagination serve you. We all live behind scrims, look through scrims; they both impede us and protect us. The true dance of the veils leads not to utter nakedness but to an ultimate coyness we do well to honor.

Dickinson's "A Charm" miniaturizes the extended action of *La Princesse de Clèves* into a brief encounter imagined and declined. Madame de Clèves lifts her veil only high enough and long enough to admit her love for the Duc de Nemours, not to act on it, even though no further social obstacle separates them. Then, for the remainder of her life, she seems to regret having made even that chaste revelation. The eight short lines of "A Charm" rely on allegorical players (the Lady) and capitalized abstractions *(Image)* and offer no dialogue, no movement more overt than *peers*. Nevertheless, two softly rhyming and punning hymn stanzas draw us into a situation and an action as persuasively human as the classically staged dramatics of the novel. Totally opposed in form and length, *La Princesse de Clèves* and "A Charm" complement each other so vividly as to appear to contain each other, to generate each other reciprocally. We have few such literary pairs.

How could these two vibrant women, fully attuned to the world around them, come to believe that fulfillment lies in renunciation, that "It was the Distance— / Was Savory—" (number 439)? There are several answers. Both were familiar with the troubadour poets and with the stories of what we now call courtly love, particularly those of Lancelot and Guinevere. From these sources came the tradition that love in its purest form entails abnegation and suffering, which ennobles all parties far more than promiscuity and pleasure. And both women confronted some form of love fright—wariness of experience, a tendency to withdraw into one's shell in order to protect a personal fantasy and enshrine a higher truth of the imagination.

Fifteen years after the rush of events and feelings that led to "A Charm," Dickinson confronted a situation curiously similar to that of *La Princesse de Clèves*. One of her father's closest friends, a prominent Massachusetts Supreme Court justice named Otis Lord, lost his wife in 1877. He had known Emily all her life; he was sixty-five, she forty-seven. They soon acknowledged to each other a deep attachment that had evidently grown over a fairly long period. Fifteen surviving letters from Dickinson to Lord reveal a wide range of feelings, including sexual passion for the man she could call "Sweet One" and "Naughty One." Judge Lord proposed marriage. Dickinson, who at twenty had sent coy valentine verses to her father's law students, who had declared "My business is to love," now had to deal with a resolute widower who did not plan to move to California. He wanted her to move in with him. The Amherst princess retreated to the nunnery of her upstairs bedroom and wrote to her lover letters that record the encounter of ardent emotion and stern constraint. She becomes both Pan and Syrinx.

> *Oh, my too beloved, save me from the idolatry which would crush us both—*
>
> *Don't you know you are happiest while I withhold and not confer—don't you know that "No" is the widest word we consign to Language?*
>
> *The "Stile" is God's—My Sweet One—for your great sake—not mine—I will not let you cross—but it is all yours, and when it is right I will lift the Bars, and lay you in the Moss—You showed me the word.*
> *I hope it has no different guise when my fingers make it. It is Anguish I long conceal from you to let you leave me, hungry, but you ask the divine Crust and that would doom the Bread.*
>
> (*Letters*, II, 617–18)

In the third, almost-steamy passage, "Stile" is not an archaic spelling of style. The word refers to the place in a fence where steps or rungs (or sometimes a turnstile) allow passage to a person and not to cattle or sheep. "You showed me the word," she writes, implying that Judge Lord first used this image of privileged access. Her response: "I will not let you cross." Yet it sounds as if they have met intimately and passionately at least over the stile afforded by searching letters such as these. What imagery could be more explicitly sensual than "I will lift the Bars, and lay you in the Moss"? Then "the divine Crust" returns to a chaste abstemiousness. The confinements Emily Dickinson imposed on herself led more to intensity and variety of feeling than to monotony.*

They never married. Lord died seven years later, in 1884.

---

*In a paragraph that deals with this correspondence, Camille Paglia allows her frequently tonic reading of Emily Dickinson to lapse into tendentiousness. "Her letters to Lord are contrived and artificial. The voice belongs to her twittering

For all her white gowns and hair pulled back in a bun, it should be evident now that Dickinson had nothing of the prude in her. From farm animals, from her sexually active sister, Vinnie, from her brother's complex marriage, and from her own daring imagination, she understood all about the nature of erotic rapture. There was no aspect of life that she shunned, that could not arouse her gift of gossip and her sense of mirth. To the end of her life, her favorite adjective was *funny*. She lived for jokes and stories and told them in her letters and—transformed—in her poems.

Free of prudishness, Dickinson's exultant abnegation contained a strong component of aestheticism. The pleasures she sought tended not toward paroxism that overwhelms the mind but toward a heightened awareness that mediates between intensity and moderation. Like the Princesse de Clèves, she strove to conserve the whole loaf of happiness rather than to consume the crusts that life usually throws our way. Their moods lie as close to epicureanism, "The Banquet of Abstemiousness," as to fear of living. I find greater strength of character and more true feeling in the roles of Syrinx and of the Princesse de Clèves and of Emily Dickinson (both in her poems and in her life) than in the roles of Don Juan and Faust. Those two fancy-grade hit-and-run drivers leave numerous victims strewn in their wake; Madame de Clèves and Dickinson seek full partners, seek lasting union, and turn away from anything that falls short.

In spite of the evidence here assembled, I do not believe that this contrast is solely the result of a difference in temperament between men and women. Women can be predators; men can show restraint. We do have to take into account, particularly in earlier periods, a difference in experience permitted to the two sexes by society. And I cannot readily cite a novel, poem, or play (*Axel* excepted) that casts a man in the role of exultant abnegation.

---

feminine personae, whom she tucks in becoming postures of devotion" (*Sexual Personae*, 670). This dismissive interpretation is required by Paglia's thesis that "the homosexual-tending Emily Dickinson" writes essentially in the personae of a sadomasochistic male, "a hierarch requiring the sexual subordination of her petitioners." What a pity that Paglia's misreading and overestimation of the Marquis de Sade and her insistent vision of Dickinson as a Sadean temperament lead her to exaggerate genuine insights into Dickinson's violent imagery and motifs of domination.

The strongest candidate comes from Flaubert's pen. *A Sentimental Education* (1869) relates the meeting of Frédéric Moreau and Madame Arnoux many years after their youthful, hopeless, and undeclared passion. Her husband is now ruined and an invalid. Still beautiful with white hair, she wears a veil. Frédéric has sunk into inertia and is surprised by her early evening visit to his apartment. During a walk, they at last confess their love. "We shall have truly loved each other." "How happy we might have been!" When they return, she removes her veil, and they embrace. Then she draws back. "I wish I could have made you happy." He wonders if she is offering herself to him.

He smokes a cigarette. Before she leaves, she cuts a long lock of hair for him. His feelings include awkwardness and a revulsion bordering on fear of incest. But most basically, he pulls back "out of prudence and in order not to degrade his ideal." The emotional charge of the scene accumulates because of the closeness of the encounter and because of the trivial items—white hair, veil, lateness of the hour, cigarette—that appear to prevent it from going any further. Their love is both thwarted and preserved. Neither of them feels exultation.

All these stories, literary and parabolic, tell us that neither promiscuity nor abnegation can escape selfishness. Most of us do not propose to live lives of such high relief. Most of us will seek, and find, a middle way. But we would do well to ponder the results for other people of promiscuity and abnegation. *Carpe diem* may not always lead to the greatest happiness for anyone.*

---

*Marcel Proust understood and described this higher epicureanism, which values imagination over satisfaction. Ultimately, he appealed to a doubling of experience in cumulative time, but his point of departure lies close to that of the Princesse de Clèves and of Emily Dickinson. "Nothing is more alien to me than to seek happiness in any immediate sensation, and even less in any material realization. A sensation, however disinterested it may be, a perfume, a shaft of light, if they are physically present, are too much in my power to make me happy" (Bibesco, 119).

Near the end of *In Search of Lost Time,* Proust goes one step further to illuminate the psychology of elective abstinence. He refers to "the inexorable law that one can imagine only what is absent" (III, 872).

In comparison to Proust's analyses, I find Beckett's stark words on the subject unjust both to himself and to Proust: "the wisdom that consists not in the satisfaction but in the ablation of desire" (*Proust*, p. 6). None of the parties here discussed, least of all Mme de La Fayette and Emily Dickinson, propose "ablation" by some psychological equivalent of surgical removal. They envision a transposition

To support her choice of renunciation, Dickinson makes a very ambitious claim in "A Charm" about the life of the mind.

Lest Interview—annul a want
That Image—satisfies—

Can imagination alone truly sustain us if reality fails? Dickinson proposes, I believe, a strategic retreat to a position that both eludes and contemplates sheer experience. From there, we can acknowledge that some precise states of body and mind are characterized by fragility; they may be shattered by too close, too rough, or too prolonged an encounter with the desired object or person. Our proverbs can be merciless on the subject. Familiarity breeds contempt. Absence makes the heart grow fonder. Dickinson looks out intensely from her special need not to approach too close to other people, to experience itself. She expresses an uncertainty principle of the heart, an indeterminacy principle of the human psyche. For her, the soul is a domain as resistant to observation and exact measurement as an electron hidden in the atom. Neither in physics nor in the life of feelings does this situation apply for ordinary events in the scale of our everyday behavior. But when Dickinson with her "Veil" and the physicist with his imaginary gamma-ray microscope move downward and inward to the tiniest order of magnitude, then they report a limit on our reach, on our knowledge.

Some thinking people bridle at the thought that any such barrier faces us anywhere. But physicists have made their peace with the indeterminacy principle. In the domain of feelings, we must not disdain the lucid resoluteness of Emily Dickinson and the Princesse de Clèves, as well as of Syrinx and of Maggie in *The Mill on the Floss*. They do not shrink from the implied paradox: that to acknowledge a limit on experience may extend our freedom to be ourselves. Not many forms of forbidden knowledge approach so close to our own lives as the prospect of abstinence and its rewards.

---

of desire to another level of experience—memory, contemplation, fantasy, reenactment, sublimation.

Chapter V

GUILT, JUSTICE, AND EMPATHY IN MELVILLE AND CAMUS

■

I hate that fatuousness of a mind that excuses what it explains . . . and that analyzes itself instead of repenting.

—Benjamin Constant, *Adolphe*, 1816, "Réponse de l'éditeur"

A two-century gap between Mme de La Fayette's novel and Emily Dickinson's poem did not prevent them from expressing a startlingly similar reluctance about approaching full emotional and physical experience—the opposite of Faust's and Frankenstein's resolve to grasp and shape experience to their own ends. Only half a century separates Melville and Camus, and I shall trace in a novel by each of them a different aspect of forbidden knowledge from any considered so far. *Billy Budd* and *The Stranger* confound us as readers (as some of the characters are confounded) by offering us information that interferes with a simple interpretation of the plot. We come to know too much about the characters to be at ease with the working out of justice under the highly strained circumstances described. As a result, the interpretation of these two short novels has provoked lengthy disputes and has led to troubling errors. Melville and Camus carry us to a further aspect of the troubling questions: Are there things we should not know? Can knowledge get in the way of justice?

1. *BILLY BUDD:* REALIST ALLEGORY

The great territorial and commercial expansion of the new American republic did not lift Melville's career as writer. After a successful start as the author of nautical adventure novels, he responded to the failure of *Moby-Dick* by appearing to retire into his New York customs-house job. But the analogies of the ship of state and the ship of individual character would not leave him. When he died in 1891, Melville had nearly finished in manuscript the short historical novel *Billy Budd, Sailor.*

Many years passed before the publication of this enigmatic narrative, partially based on a mutiny in 1842 aboard the U.S. Navy brig *Somers*. Melville's much-admired cousin, Guert Gansevoort, was first lieutenant on the *Somers* at the time. Melville himself was twenty-three, whaling and jumping ship in the Pacific. Forty years later, in the 1880s, the *Somers* affair was still being written about when Melville began work on a related poem, "Billy in the Darbies." What we read as a consecutive novel is essentially the extended, hypertrophied headnote for that poem. Its thirty-two lines of rough but not free verse maintain their place at the end of a slow-starting story that climaxes in thirty pages of intense action. A synopsis offers both an X-ray version and a caricature of the story.

Impressed into service aboard a British naval ship during wartime, the Handsome Sailor, Billy Budd, conducts himself with unaffected "natural regality" comparable to that of Adam before the Fall. Only an occasional organic defect or stutter flaws his demeanor. Billy unwittingly arouses the passionate envy and perhaps the desire of Claggart, master-at-arms, an intelligent petty officer mysteriously associated with "natural depravity."

In front of Captain Vere, a just and undemonstrative disciplinarian, Claggart accuses Billy of fomenting mutiny. Rendered speechless by the false charges, Billy strikes and kills Claggart. Though Captain Vere seems almost "unhinged" by this "mystery of iniquity," he sets in motion a three-man drumhead court. After a deeply troubled discussion led by the captain, the court swiftly condemns Billy to be hanged from the yardarm. Billy's execution, marked by his last words, "God bless Captain Vere!" echoes the Crucifixion and survives in two garbled versions: the official naval chronicle of a villainous Billy and a popular ballad depicting a suffering, noble Billy, "Billy in the Darbies."

As in *Moby-Dick*, layers of nautical detail establish a foreground of convincing realism. Recurrent biblical allusions to Adam, Satan, Abraham, and Joseph engage the narrative at critical moments in allegory. Two key passages reveal how steadily the tale moves from realism toward allegory. The first passage comes at the close of a long characterization of Claggart.

> *Now something such an one was Claggart, in whom was the mania of an evil nature, not engendered by vicious training or corrupting books or licentious living, but born with him and innate, in short "a depravity according to nature."*
>
> *Dark sayings are these, some will say. But why? Is it because they somewhat savor of Holy Writ in its phrase "mystery of iniquity"? If they do, such savor was far enough from being intended, for little will it commend these passages to many a reader of today.*
>
> (Chapter 11)

The tentative phrasing of the two paragraphs only underlines their portentousness. The phrase "mystery of iniquity" (II Thessalonians 2:7) designates the problem of the existence of evil in a God-created world, the problem addressed by Leibnitz with the modern term *theodicy*—Milton's avowed subject in *Paradise Lost.* What absorbs Melville is how Claggart's evil infects the innocent Billy through an obscure causation we call fate.

The second key passage presents that infection or moral reversal as tragic and inevitable. It takes an attentive reading to follow the paradoxes and reversals described in these sentences.

> *In the jugglery of circumstances preceding and attending the event on board the* Bellipotent, *and in the light of that martial code whereby it was formally to be judged, innocence and guilt personified in Claggart and Budd in effect changed places. In a legal view the apparent victim of the tragedy was he who had sought to victimize a man blameless; and the indisputable deed of the latter, navally regarded, constituted the most heinous of military crimes. Yet more. The essen-*

> *tial right and wrong involved in the matter, the clearer that might be, so much the worse for the responsibility of a loyal commander, inasmuch as he was not authorized to determine the matter on that primitive basis.*
>
> (CHAPTER 22)

Notice that all three characters are arrayed together in this tense passage. It tempts us to make a double allegorical leap. The *Bellipotent* is a ship of war that represents not only a ship of state in crisis but also the ship of a divided and unified individual: Claggart as evil, Billy as good or innocence, Captain Vere as the authority of reason trying to maintain order. The tragedy is not lowly Billy's, but Vere's; the captain dies years later muttering Billy's name. As a wartime commander, he is duty-bound to judge the killing according to the forms of naval justice. The principle is not new to him. " 'For mankind,' he would say, 'forms, measured forms, are everything' " (Chapter 27). Tocqueville uses the same word—*forms*—to designate the traditions and customs he finds lacking in an open, democratic society. The forms of naval justice are barely adequate to deal with the mystery of iniquity, with the instantaneous blow that transforms Billy into a murderer and Claggart into a half-innocent victim.

The end of the tale offers us two contradictory versions of the events. In the official naval records, ordinary seaman Billy Budd becomes a knife-wielding alien and Petty Officer Claggart a discreet, respectable gentleman. In the popular ballad "Billy in the Darbies"—that is, the poem that generated the whole story in Melville's imagination—the sailor in irons dreams of his death as a form of sleep. These versions correspond loosely to the two critical interpretations the novel has inspired, one or the other usually quoted on the cover: "Melville's Quarrel with God" or "Melville's Testament of Acceptance." It is essential to point out that the novel does *not* say to choose one interpretation or the other. Even Captain Vere, whom it is easy to see as an inflexible, unsympathetic martinet, knows that the situation is more complex than his official conduct can acknowledge. This father figure partially represents Melville trying to come to terms with the loss of his two sons, one to suicide, one to sickness. More profoundly, Vere represents the attempt of an upright and intelligent man to come to terms with the intellectual currents of the nineteenth century: irreligion, science, evolution, democracy. The circumstances of Billy's court-martial tax Vere to the limit. Because Vere accepts full responsibility for his ship, his dilemmas carry the themes of forbidden knowledge a step further than in *Paradise Lost* and *Faust.* Vere has no God figure to help him.

Billy Budd's probing of unexpressed states of mind may well be what Melville wished to designate with the puzzling subtitle he penciled into the margin of the manuscript: "An Inside Narrative." It doesn't refer to any form of narrative omniscience; many crucial events remain unknown even to the faceless narrator. For us, in the present discussion, the subtitle looks forward to another novel written fifty years later.

2. *The Stranger:* An Inside Narrative

A Frenchman born and brought up poor and fatherless in Algeria, Albert Camus worked there as journalist and dramatist before becoming a major Resistance figure in continental France during World War II. His most haunting work, a short novel called *The Stranger* (1942), was immediately singled out as a major exhibit of existentialism. The philosophical and literary movement, which swept Europe for two decades after the war, enshrined Camus as a major hero opposite Sartre, with whom he broke in 1952 over the latter's Soviet loyalties. Camus received the Nobel Prize in 1957 and died in 1960 in an automobile accident.

The Stranger has remained an astonishingly timely book for the last half of the twentieth century. The novel's popularity reached its peak in the 1960s and contributed to the formation of the cool, hip, detached hero, or antihero, of that decade. There were a few copycat crimes on beaches in California and elsewhere. The book's idealized affectlessness costumed itself in such words as the *absurd, authenticity,* and *sincerity.* As the century closes, *The Stranger* is still widely read and equally widely *mis*read. I link that serious misreading to an aspect of forbidden knowledge and will compare it to the closely related action of *Billy Budd.*

Melville's subtitle "an inside narrative" fits Camus' *The Stranger* like a glove. The first half of Camus' tale confines the reader inside a single intermittently vivid yet numbing sensibility. One diffident character tells us his own story. Then during the interrogation and trial scenes forming the second half, everything happens all over again in retrospect, according to a terse principle Camus affirmed at the same period: "To create is to live twice" *(The Myth of Sisyphus).* My synopsis provides an "outside" narrative of the action.

A self-effacing French office employee in Algiers, Meursault, writes in a curiously neutral yet graphic style. He relates the events of his mother's wake and burial, how he picks up a new girlfriend, Marie, the next day, and how he gets embroiled in the unsavory squabbles of Raymond, a pimp who lives in the same apartment building. Meursault takes a trip to the beach with Marie and Raymond and drifts inertly into a sinister confrontation with Raymond's Arab enemy. In a demented or exalted moment on the burning sand, goaded by the sun and blocked by the Arab from reaching the cool spring, Meursault kills the man with five shots from Raymond's pistol.

During the investigation and trial, all these events are replayed for Meursault as both monstrous and model, bizarre and natural. He finally accepts his guilt and adjusts to the monotony and deprivation of existence in prison while awaiting his execution. Near the end, Meursault lashes out violently at the prison chaplain for trying to divert him from the only two courses left open to him as a man: to live peacefully with his sensations and his memories, and to die defiantly by the guillotine, thus affirming his life.

Anyone who has read this troubling modern classic probably remembers the pervading moral deadness of Meursault's life and character punctuated by moments of intense physical immediacy. It helps very little to attach the labels "absurd" and "alienated" to his existence. One perceptive commentator has noticed how closely Meursault's behavior parallels the automatism that Bergson in his essay *Laughter* identifies as the source of humor. In an interview, Camus mentioned humor as the theme most neglected in his work. A more likely reference or even source for Meursault's obtuse sensuousness comes from a philosopher whose work Camus was reading in 1938–1939 while working on *The Stranger.* "So the animal lives *unhistorically.* . . . It merges entirely into the present, it knows nothing of dissembling, hides nothing, and seems always exactly what it is, and so cannot help being honest" (*Unmodern Observations*, 170, tr. Gary Brown).

Nietzsche's description of animal consciousness follows close behind Rousseau's *Discourse on the Origin of Inequality*: "I almost dare to affirm that the state of reflection is a state contrary to nature and that the man who meditates is a depraved animal." Camus catches the mood of Meursault's stunted mind in great part through a muted style tending toward disfunction and parataxis. Nothing connects. Things just happen. In a famous commentary, Sartre calls our attention to the indolence and indifference of this dumb writing. Camus carries his flat, chopped-off style a step further than Kafka and Hemingway. It conveys the metaphysical drift of our age as acutely as the montage principle in cinema and in painting. *The Stranger* offers us the inside narrative of virtually vacant mental states verging on autism.

Yet Camus depicts Meursault's animal consciousness being pushed little by little toward self-awareness. The process begins after his mother's burial, after Marie spends Saturday night with him, at the end of the long, idle Sunday that follows.

> *I wanted to smoke a cigarette at the window, but the air was getting colder and I felt a little chilled. I shut my windows, and as I was coming back I glanced at the mirror and saw a corner of my table with my alcohol lamp next to some pieces of bread. It occurred to me that anyway one more Sunday was over, that Maman was buried now, that I was going back to work, and that, really, nothing had changed.*
>
> (24, tr. Matthew Ward)

Meursault does not yet see *himself* in the mirror. He glimpses only a few fragments of his environment. He seems to register for a moment the utter vacancy of his life. A variety of scenes builds on this one. Meursault vaguely senses that the robotlike woman in the restaurant, who scrupulously writes out her own check with tip, embodies a caricature of himself. In prison, he studies the reflection of his face in the tin pannikin and realizes that he has been talking to himself. In the courtroom, a young journalist gazes so hard at

Meursault that he gets the impression he's "being scrutinized by myself." Later, while Meursault awaits execution, the prison chaplain visits him and gazes at him constantly during a lengthy and antagonistic conversation. Finally, the chaplain's insistent words—"I'm on your side. I'll pray for you"—provoke Meursault into seizing him and yelling at him. At this point, Meursault seems finally to perceive himself, to take hold of himself. "Nothing is important," he shouts at the chaplain. "Life is absurd." Meursault has glimpsed in the succession of reflections something that inspires defiance in him, followed by "tender indifference" when he is alone again. This sudden surge and fall of feeling occurs in the last three pages. One could read the final sentence of *The Stranger* as an ironic version of the execution scene at the end of *Billy Budd* (Billy shouts "God bless Captain Vere!"), even as a parody of that scene.

> *For everything to be consummated; for me to feel less alone, I had only to wish that there would be a large crowd of spectators the day of my execution and that they would greet me with cries of hate.*
>
> (123)

3. Comparing Two Specimens

But Camus didn't read *Billy Budd* until after he had written *The Stranger*. Then in an encyclopedia article, he praised Melville's novel and wondered whether Billy's death represents a protest against a blasphemous violation of human justice, or a resigned assent to the terrible order of Providence. Those are the two standard interpretations already mentioned. We have seen Providence before in story after story; it is one of the guises of forbidden knowledge. It is surprising that Camus did not refer to the arresting similarity between *Billy Budd* and *The Stranger*. They are like two mineral specimens, different in color and texture, yet whose crystalline structures resemble each other. The two narratives turn on essentially the same situation. From one point of view, a violent, lethal deed—not of passion or premeditation, but of impulse—is described from inside as an innocent act. From an opposed point of view, a rigid system of justice finds the same deed to be criminal enough to merit the death penalty. Neither man defends himself against the charges. Neither man feels remorse or moral anguish, even though each accepts his guilt. It is, I believe, this sustained moral ambivalence that makes it difficult to deal adequately with the two books.*

As Camus points out in his encyclopedia article on Melville, *Billy Budd* presents its moral dilemma with the starkness of a classic tragedy. Confronted by the two nearly stereotypical figures of good and evil in Billy and Claggart, Captain Vere presents to his fellow officers sitting in judgment a case against Billy that he supports with the responsibilities of his rank. Order must be maintained, even in the face of his own inclinations to favor Handsome Billy. In *The Stranger*, the place of Captain Vere is occupied not by any corresponding character, not by the three judges and the jury, but by the reader. It is a vast difference. The reader must decide between Meursault's seemingly candid inside account of how the awful events somehow produced themselves through no fault of his, and the prosecutor's wandering and sometimes odiously righteous account of Meursault's criminal behavior. Still, the prosecutor shows Meursault to be unwilling to face the consequences of his acts and bereft of moral awareness. The jury remains remote; its decision barely concerns us as we contemplate the original events and the trial and make up our own mind about them.

Writing about *Billy Budd* in *Beyond Culture*, Lionel Trilling felt the need to report that most of his hundreds of students over the years condemned Captain Vere for condemning Billy Budd. After teaching *The Stranger* off and on for thirty years, I must report a

---

*In the opening section ("At Sea") of *Quatre-vingt-treize*, Victor Hugo narrates an exciting scene leading to a comparable yet very different moral dilemma. The negligence of a cannoneer aboard a naval vessel in wartime allows a cannon to tear loose from its lashings, to crash across the decks, causing massive damage and threatening to sink the ship. The same cannoneer courageously catches and resecures the cannon, thus saving the ship. The newly appointed commanding general on board first decorates the cannoneer for valor and then orders that he be shot for negligence in combat. The crew grumbles and carries out the order. The cannoneer makes no protest. The general has proved his mettle by dispensing justice as inflexible as Captain Vere's. And the story has barely begun.

similar response among students to Meursault and to the guilty verdict pronounced against him. At first, I partly agreed with the students. Later, after shifting my position, I began keeping track of their reactions. In 1975, on the midterm examination of an undergraduate French course of twenty-two students, I asked whether the prosecutor was justified in calling Meursault's behavior "monstrous." Though written in rudimentary French, many answers seem remarkably eloquent.

> *Meursault was wrong to kill the Arab. But he didn't mean to. The death was almost an accident. He's not a criminal. He's only a man in a bad situation.*

> *I think that the real monster, the person who cannot control his emotions, is more the prosecutor than Meursault.*

> *One must understand Meursault in order to realize that his actions were not his choice but simply what happened.*

Eighteen students out of twenty-two sympathized with Meursault, called for "understanding" his situation, and defended his behavior as primarily "different."

In 1990, in a graduate-undergraduate comparative literature course of thirty students, I asked for a synopsis (in English) of *The Stranger* and a brief commentary on the action. Many of the papers tapped reserves of genuine passion.

> *Meursault is sentenced to be decapitated more for the person he is than for the crime he has committed.*

> *Meursault sees objectively and impersonally . . . and learns to live as Job did, without judging life by human standards,* [and thus] *transcending anthropomorphism.*

> *Reading Camus'* The Stranger *is a bit like witnessing a swimmer struggling against an overpowering current. The swimmer, who maintains no pretenses and clings to no false hopes, is challenged head-on by a society that prides itself on "meaningful" conventions and order. . . . Just as the Stranger yanks away the security blankets*

> *civilization tenaciously clings to, so does he jolt the reader to a new level of consciousness.*
>
> [The novel presents] *the French judicial system's inability to cope with Meursault, whose honesty leads him to be sentenced to death.*
>
> *The tragic protagonist, Meursault, stoically narrates his existence misunderstood by a judgmental French society.*

These comments, all of which deserve careful attention, testify to a grave misreading leading to moral myopia. In most of them, the basic fact of the murder is discounted, not mentioned, virtually overlooked. They assume that Meursault has told his story honestly and sincerely. What more can we want? The second group of students had read *Billy Budd* earlier and had discussed the tragic necessity of Captain Vere's drumhead court and its verdict. (Many questioned, very properly, the need for summary execution.) Now, when asked to assume the role of Captain Vere in facing a comparable situation, many of them capitulated to the voice of Meursault narrating his own tale. Camus created a cool, flat, artificially natural style for most of the episodes. Set against that monotonous landscape, the semiritualistic killing on the beach releases in Meursault a glorious burst of lyric intensity. The murder scene combines the crescendos of a gratuitous act and an epiphany. The seizure of that moment apparently makes Meursault lose consciousness and suffer memory loss. His later attitudes and behavior remain mysterious because we are never given a full account of what happened right after the murder and of how he was apprehended.

Camus' narrative has the power of magic incantation in modern dress. It makes one forget that Meursault never thinks of or refers to the human being he has killed. He experiences no regret, no remorse. To the examining magistrate, Meursault identifies his feelings about the deed as *ennui*—vexation or annoyance. On the last page, by saying "I felt ready to live it all over again," he appears to reaffirm his crime and his punishment as the only source of his identity, as his signature.

Is this laconic murderer our modern Prometheus? Can any person so disingenuous, so unambitious, and so unassuming as Meursault possibly be a monster? Camus himself compounds the difficulty in the brief preface he wrote in 1955 for an American textbook edition.

> *Sometime ago I summed up* The Stranger *in a sentence, which I grant is very paradoxical: "In our society any man who doesn't cry at his mother's burial runs the risk of being condemned to death." I meant simply that the book's hero is condemned because he doesn't play the game. . . . He refuses to lie.*
>
> . . .
>
> *For me Meursault is no derelict but a poor naked man in love with a sun that leaves no shadows. Far from lacking feeling, he is animated by a profound passion—profound because it remains mute, a passion for the absolute and for the truth.*
>
> . . .
>
> *I have also gone so far as to say, paradoxically once again, that I tried to present in Meursault the only Christ we deserve. It should be clear from my remarks that I intend no blasphemy and speak only with the slightly ironic affection an artist has the right to feel toward the characters he has created.*

These astonishing claims by the author of *The Stranger* have rarely been challenged. Originally, Camus said in his Nobel Prize acceptance speech, he conceived Meursault as a figure of "negation." He allowed a haunting ambiguity to hover over his laconic hero. Thirteen years later, in the passage quoted above, Camus presents Meursault as a hero of nonconformity and uncompromising truth.* Is that how we should read *The Stranger*? Has Camus forgotten that Meursault lies at least twice for Raymond: once in writing the letter to the Moorish woman who, Raymond claimed, had cheated on him; and once to the police? Was Meursault condemned to death for refusing to lie and to play the game of saying more than

---

*René Girard's essay "Camus's Stranger Retried" makes a strong case for Camus having written in *The Fall* (1956) a thorough-going rebuttal of "the implicit indictment of the judges" expressed by *The Stranger*. I fully concur with Girard's description of how Camus attempts in *The Stranger* to narrate a crime without a criminal. The puzzle lies in how the 1955 preface could collapse the original ambiguity into a Christ figure, whereas first *The Rebel* (1951) and then *The Fall* distance themselves increasingly from this Romantic myth of the persecuted Self.

he feels? Or was he guilty of letting himself be drawn into settling a score for a violent small-time pimp and of killing an Arab? Like the students I have quoted, Camus insists in his preface that Meursault is condemned for his sincerity. Camus conveniently overlooks the fact that his hero committed murder. I believe that Camus' preface provides a case of an author who grievously misunderstands his own work and his most famous character. Possibly his repetition of "paradoxical" and his use of "ironic" in the last sentence should lead us to reverse the meanings of "hero" and "Christ." But I don't think so. Camus' wandering yet succinct prose in his novel seems to have hypnotized him along with his readers.

We should be closer now to perceiving the paradox of *The Stranger*. How do we explain our spontaneous or perverse admiration for a generally morose citizen who is duped into murder and experiences no remorse? In comparison, Adam and Eve and even Faust display greater responsibility for the consequences of their deeds. In Part I, Meursault makes no effort to hide his feelings, if he can find any. He remains insensitive to his own actions and to their consequences for others. Though his new girlfriend, Marie, has a name and sometimes occupies his thoughts, she barely concerns him as a person, any more than does his nameless victim, the Arab. Meursault is self-absorbed rather than self-conscious and describes tiny sensations of eating, waiting, smoking, and watching—describes them so vividly that we are drawn into his vacuous life. Then, yielding to the gradually accelerating tempo of the style, we live through the whole sun-spangled, heat-driven scene on the beach *from the inside.* It assumes the monstrous form of sheer accident enacted as inexorable fate. A slow accumulation of fragmentary sensations absorbs us into a mind that does not draw back from—from what? From letting his "whole being tense up" in such a way as to make him squeeze the trigger of the loaded gun that just happens to be in his hand. But the deafening report of the shot wakes him at last out of his zombie-like existence. At that moment, "everything began," and Meursault "understood" what had happened, what he had done. Instead of recoiling, he now affirms his half-conscious deed by deliberately firing four more shots into the dead Arab. The sleepwalker becomes a criminal who feels exhilaration. He does not become a human being appalled at the spectacle of murder. The shock of a criminal act gives Meursault his first startling experience of being fully alive.

"It requires a considerable effort on [the reader's] part to disengage himself from the rhetoric of the story to the extent of recognizing something monstrous" in the principal character. What Denis Donoghue says in *Thieves of Fire* about D. H. Lawrence's story "The Captain's Doll" applies accurately to readers of *The Stranger*. Most readers accept their first sympathetic response to poor, unambitious, victimized Meursault. The seemingly artless way he tells his own story disarms our ability to detect an unreliable narrator. *The Stranger* offers the most convincing version ever written, I would say, of the sincerity plea made in exoneration of an incontrovertibly criminal action. The rhetoric apparently deceived Camus himself a decade later. Meursault's "sincerity" in Part I lies close to pathological autism. In Part II, during the extended process of waking up to himself as a responsible person, Meursault yearns both to revert to the soulless existence of Part I and to dismiss the perfectly justified guilty verdict by defying it. Both responses constitute a lie to himself as a potential human being.*

I cannot help seeing this miniature novel as a parable, a piece of subtle didactic writing whose meaning reveals itself gradually to those who read carefully. But because of the subtle blandishments of the inside narrative, which seduce many readers into empathizing with a criminal, the parable misfires. The moral lesson—that no existence can be called human that does not accept a minimum of responsibility for itself, for its actions, and for others—is too easily overlooked. I remain astonished at the extent of misunderstanding and distortion that Camus expressed in his 1955 preface. Fortunately, it is not included in regular trade editions either in French or in English.

---

*Robert C. Solomon has written a careful treatment of the themes of lying and self-consciousness in Meursault.

4. Understanding, Blaming, Forgiving, Pardoning

I believe a proverb will help explain why so many readers of *The Stranger* are led into seeing a monstrous criminal as a hero of authenticity, and why the dilemma of Captain Vere in *Billy Budd* develops a very different hold on us in a comparable situation. The proverb that comes to hand here has wide familiarity. Yet it appears in only one standard collection. Its generic form appears to be French: "*Tout comprendre c'est tout pardonner.*" We translate it into English tersely, dropping the *tout*: "To understand is to forgive."* Some modern variants add local color and alliteration. "Never criticize anyone till you've walked a mile in his moccasins." The poet Henri Michaux included this cautionary pastoral version in one of his collections: "If the wolf understands the sheep, he'll die of hunger." However phrased, these distillations of folk wisdom deal with the power of empathy to sway our judgment.

So arresting a formula as "*Tout comprendre c'est tout pardonner*" could not long exist without generating its polar opposite. La Rochefoucauld provides a subtle version. "If the world were aware of the motives behind them, we would often be ashamed of our finest actions." In other words: "To understand is to condemn." G. B. Shaw says so without flinching: "If a great man could make us understand him, we should hang him." We could recast it in neutral terms: Full understanding compels full judgment. But when do we ever reach full understanding?

I single out "*Tout comprendre c'est tout pardonner*" in part because the proverb links our tendency to heroize Meursault to the spell of Camus' seemingly transparent narrative style. That style makes us believe we understand Meursault. But the proverb also encapsulates a variant of moral relativism. Travelers have always noticed that customs and laws can be very different on the other side of a frontier. Montaigne condoned cannibalism among South American Indians, not in his native Bordeaux. Not until modern times has

---

**The Home Book of Proverbs* cites Mme de Staël's *Corinne* (1807) as its first source ("*Tout comprendre rend très indulgent*"), followed by references to Tolstoy's *War and Peace* (I, 1, 26) and Unamuno's *Essays and Soliloquies*. But the last reference, to a German proverb, is evidently the oldest. "*Ein Ding ist nicht bos wenn man gut es versteckt.*" In other collections one can find: "*Peché avoué est à moitié pardonné.*" "A sin confessed is half forgiven."

relativism been widely applied *within* a culture. In the opening pages of *Diary of a Writer*, Dostoyevsky editorializes about the "acquittal mania" that is affecting juries across Russia in the 1870s. Juries see criminals as victims of circumstance. "Who is guilty? The environment is guilty . . . there are no crimes at all." Though Nietzsche found ways to justify crime and immorality among the strong, he could not tolerate sympathy for misconduct among the weak.

> *One knows the kind of human being who has fallen in love with the motto,* tout comprendre c'est tout pardonner. *It is weak. . . . It is the philosophy of disappointment that wraps itself so humanely in pity and looks sweet.*
>
> (*The Will to Power*, 81)

As a young man, Marcel Proust twice filled out a questionnaire that included the following item: "For what faults do you have the greatest indulgence?" At age twelve Proust answered: "For the private life of geniuses." At age seventeen his response was: "For those I understand." In Robert Musil's immense novel *The Man Without Qualities*, the character Moosbrugger, rapist and murderer, becomes the darling of intellectuals who admire his forthright testimony and find reasons to exculpate his behavior. Moosbrugger begins to look like a wily jovial version of Meursault. Thomas Mann sets the scene for the inside narrative of *Death in Venice* by alerting us to its moral content. After a long period of doubt and antisocial thinking, Aschenbach has returned "from every moral skepticism" to a more balanced view of individual responsibility. This change is described as "the counter move to the laxity of the sympathetic principle, that to understand all is to forgive all." A popular New England folksinger, Banjo Dan, often performs a long ballad called "Werewolf." Each stanza recounts further horrors committed by the Werewolf, followed by this refrain.

> *He's ravished a few maidens,*
> *He drank the blood of many poor children.*
> *But if you knew him you'd see*
> *The Werewolf is like you and me.*

Every one of these authors recognizes the empathy-sincerity plea: Anyone in the same situation would do the same thing. Proust and Banjo Dan seem to welcome it. Dostoyevsky, Nietzsche, Musil, and Mann resist it. One cannot easily find the origin of this willingness to abdicate moral judgment, which seems to have grown throughout the twentieth century. In the West, relativism received impetus from the Enlightenment challenge to Christian morality. It was then that Lessing and Goethe calmly and cavalierly offered a new model of Faust. The personage formerly condemned to Hell, after a lengthily documented life of presumed "striving" that causes many deaths and much grief and destruction, now floats comfortably up to Heaven at God's express command. The Lord, especially in "The Prologue in Heaven," seems to "understand" Faust all too well and to forgive him in advance.

A discussion of the lures and perils of relativism could send us on a lengthy journey. Let me return, instead, to the generic proverb in order to examine its terms and its structure. "*Tout comprendre c'est tout pardonner.*" "To understand is to forgive." Composed of two infinitives (in English, in the absolute form) and the rudimentary copula *is,* the proverb takes the schematic form of a logical proposition, even of a mathematical equation. But we quickly realize that it applies not to everything in the universe but only to human actions, particularly to wrongful and evil actions.

To understand—this infinitive implies many things. First, as affirmed by Terence and Montaigne and Vico, it implies that each of us contains the whole human condition in potentiality, reaching to the furthest extremes of virtue and monstrosity, altruism and autism. Second, the infinitive *to understand* implies that we all have in varying degrees the capacity to explore that range of moods and behaviors. We call that capacity for mental exploration and experimentation *imagination,* as if it were a faculty, almost an organ. Imagination is obviously a highly complex process, yet as essentially human as primary processes like feeling and reason. Third, the infinitive *to understand* often implies that when imagination seems to carry us convincingly into another mind by empathy, we tend to interpret that person's behavior as caused by some form of fate or determination. In the twentieth century, we may choose between an exterior fate contained in society, culture, and the environment, and an interior fate—either genetic inheritance or the unconscious. To "understand" someone's behavior in this sense means to attribute it to a set of causes and to remove it from the domain of choice and free will. That form of understanding denies individual agency and responsibility for one's actions. Under such circumstances, not much remains to forgive. Fourth, *to understand* may also mean a mental operation not of empathy but of detachment and measured judgment. We seek such understanding in order to obtain a fair trial. Here, any inside narrative is subject to correction by the outside narrative of other witnesses. But the received interpretation of the proverb sets aside this meaning of *to understand* in favor of the previous meaning of empathy, of entering another person's consciousness.

After "to understand," we have another pair of words to examine: *to pardon* and *to forgive.* In common speech, we barely distinguish between them. The French generally use one verb, *pardonner,* to cover all the ground. "*Mon Père, pardonnez-leur, car ils ne savent pas ce qu'ils font*" (Luke 23:34). But the English for Christ's words on the cross could never be "Pardon them, for they know not what they do." When we translate the French proverb as "To understand is to forgive," we have made the right choice for English. *To forgive* supposes an act of imaginative empathy toward a fellow human being. *To pardon* engages the system of justice. To clarify these nuances, we need to look at a cluster of terms the English language offers us in this context. I ask for the reader's patience in an attempt to deal more precisely with words than is always necessary.

The infinitives *to exonerate* and *to exculpate* mean to clear someone of a charge, to determine that there is no offense and therefore no guilt to absolve. Two further infinitives have a more restricted meaning. *To pardon* means to remit the punishment or penalty for an offense—and by implication to recognize guilt for the offense. (By pardoning ex-President Nixon in advance of any impeachment proceedings, President Ford also established a presumption of guilt.) *To forgive* grants remission of guilt for an offense and of the resentment it may entail. The punishment is not suspended. "Starry" Vere as a man could forgive Billy for his explosive response to Claggart's lie; Captain Vere could not pardon the sailor under his command. To pardon designates a legal act; to forgive designates a moral response. After conviction for a tort, or wrongdoing, legally one discharges the penalty unless pardoned. Morally, whether forgiven or not, one is called upon to repent and do penance. Too often today we entirely overlook the last moral duty.

These distinctions among loosely used terms yield a schematic outline of five possible outcomes, *legal and moral*, when a person is brought to trial for an alleged crime.

1. Acquittal: no grounds for guilt or punishment
2. Conviction: sentencing, punishment exacted
3. Conviction and pardon: guilt maintained, punishment remitted
4. Conviction and forgiveness: guilt absolved, punishment maintained
5. Crime without a criminal: no guilt, no punishment (empathy-sincerity plea)

These outcomes allow us to locate *Billy Budd* and *The Stranger* along this sequence derived from the proverb "To understand is to forgive." *Billy Budd* fits neatly enough into the fourth category. Because of attenuating circumstances and of his "nobility" of character, all parties forgive Billy yet acknowledge the need for maintaining the severity of the punishment. *The Stranger* poses knottier problems. Outwardly the novel conforms to the second item. At the end, Meursault has been convicted and is awaiting execution. The inside narrative strives to make a case for the first item. But since neither the evidence nor Meursault's confession will justify acquittal, the first-person narrator patiently builds up the psychological climate for outcome number five. Meursault appears to be telling the simple truth about himself. The empathy-sincerity plea begins to blur all distinctions. As readers, we are drawn so vividly into Meursault's empty world and affectless consciousness that notions of wrongdoing and guilt fade away as quickly as the smoke from Meursault's cigarettes. Meanwhile, the details of the story keep suggesting that Meursault is merely human, all too human. A similar defense was offered for crucial contributions to the Nazi enterprise by a cultured architect with an immense talent for industrial organization. In *Inside the Third Reich* (1970), Albert Speer argues passionately yet contritely that he was simply carried along by the demands made on his capacities. "Completely under the sway of Hitler, I was henceforth possessed by my work. Nothing else mattered" (32). This highly placed Meursault never looked directly at the abominations he was helping to commit. And he, too, carried many readers along with him because of his seemingly candid style.

I am proposing that the students who saw Meursault as "honest" and "misunderstood by a judgmental French society" fell into a serious error partially illuminated by the proverb "To understand is to forgive." G. K. Chesterton called this attitude "the devil's sentimentality." Under carefully controlled conditions, as in listening to the "sincere" and seductive narrative voice of *The Stranger*, our empathy for another person can be stretched very far. We can venture too close and lose our perspective on humanity. Once we understand another life by entering it, by seeing it from inside, we may both pardon and forgive a criminal action. We may not even recognize it as criminal. We are all guilty in some way. How can we ever judge anyone else, punish anyone else?

That line of thinking leads to an unacceptable dilemma. Either justice is impossible and escapes us, or justice, if we do attempt to establish it, is inhuman. The action of *Billy Budd* confronts and blocks such slack thinking. Captain Vere in his fanatic resolve to maintain strict discipline aboard ship remains fully human, and tragic.* But a failure of humanity and of judgment afflicts the

---

*During the past twenty years, the most probing commentaries on *Billy Budd* have been written by legal scholars. The plain undisputed facts of Billy's striking and killing Claggart become entangled with several different bodies of law.

The incident thus lends itself to conflicting interpretations and adjudications. In contrast, Kurosawa's classic film *Rashomon* (1950), about a reported incident of rape and murder, centers not on the applicable law but on the elusive facts of the case, on the nature of truth. The most strenuous of the law-review articles on *Billy Budd,* Richard Weisberg's "How Judges Speak," reads the novel in a manner opposite to what I have argued and then uses the alleged malevolence of Captain Vere as a means of attacking a Supreme Court decision written by Justice Rehnquist. Weisberg presents Vere as reenacting at a higher level Claggart's role of ambitious dissimulator. As Claggart envies and resents the Handsome Sailor in Billy, Vere envies and resents the heroic open leadership of Admiral Nelson. Then Vere discharges that venom onto Billy, whose execution he justifies with legal deviousness and high rhetoric. Weisberg's last paragraph paints Vere as the symbolic ancestor of Stalin and Hitler.

I find the role Weisberg assigns to Nelson not adequately borne out by Melville's scrupulously written narrative. Weisberg's exposé of Vere's hidden motives

reader who overlooks Meursault's obscurely motivated murder of an Arab and who finds that Meursault's flat account of the details of his everyday life redeems the rest of his conduct. How could such an ordinary and unassuming person be a murderer? For a time, Camus himself became one of those misguided readers. He seems to have forgotten what the Greek Oresteian Trilogy, a set of tumultuous plays clearly related to *Billy Budd* and *The Stranger*, sets before us: that we cannot survive without a system of justice. We have a duty to judge and to punish crimes. (The death penalty is another question.) Orestes is not *pardoned*. He is finally *forgiven* the bloodguilt of matricide, but only after lengthy suffering at the hands of the Furies, genuine penance, and ritual cleansing. Justice is done, and a precedent is set. Two millennia later, Benjamin Constant, a contemporary of Mary Shelley, expressed (see the epigraph for this chapter) a proper scorn for explanations, analyses, and excuses for reprehensible actions. Constant favored repentance.

The trials of Billy and of Meursault take place at a great moral distance from any potential "greatness in evil" as contemplated by La Rochfoucauld, Pascal, and Goethe (see page 106). These two lowly men have not been infected with presumption and *pleonexia*. Their opposite fault lies in lack of imagination about themselves and others—in Meursault's case, affectlessness to the verge of autism.

And now we have arrived unexpectedly at a highly disputed crossroads called "the banality of evil." Like Eichmann on trial for mass murder, Meursault serves to illustrate that challenging phrase dropped in the last sentence of Hannah Arendt's book on Eichmann. Later, over and over again, she had to explain what she meant by the words.

> *. . . this new type of criminal commits his crime under circumstances that make it well-nigh impossible for him to know or feel that he is doing wrong.*
>
> (EPILOGUE, *EICHMANN IN JERUSALEM*)

---

and dark ambition comes to sound like an exemplification of the counter proverb: To understand is to condemn. More balanced treatments of *Billy Budd* have been written by Robert Cover and Richard A. Posner.

> *. . . when I speak of the banality of evil, I do so on a strictly factual level. . . . Eichmann was not Iago and not Macbeth. . . . He* merely, *to put the matter colloquially,* never realized what he was doing . . . *this lack of imagination* . . . [this] *sheer thoughtlessness . . . can wreak more havoc than all the evil instincts taken together.*
>
> (POSTSCRIPT, *EICHMANN IN JERUSALEM*)

> [The purpose of the Eichmann book was] *to destroy the legend of the greatness of evil, of the demonic force.*
>
> (INTERVIEW, *NEW YORK REVIEW OF BOOKS,* October 26, 1978)

Billy did not intend to kill Claggart; the simple sailor did not know his own strength. If there is any greatness in Melville's novel, it resides in Captain Vere's struggle. Meursault appears to have no intentions at all; he lets himself be carried to catastrophe by a wave of circumstances. I find no greatness of mind or action in *The Stranger.* Camus's two powerful crescendos of narrative-descriptive style depict a man indifferent to good and evil losing control of himself because of the banality of his imagination.*

A more substantial novel than the two we are discussing also explores this literary and moral question of dealing with a crime seen from inside. Dostoyevsky originally sketched out *Crime and Punishment* in the first person; the final third-person version remains very close to Raskolnikov and often enters his thoughts, feelings, and dreams. We frequently identify with Raskolnikov and may even feel the lure of nihilistic egoism that seethes beneath his decency and idealism. But Dostoyevsky supplies other characters

---

*I believe that "the banality of evil" and Meursault's story afford an illumination of one of Plato's troubling notions in Book II of *The Republic*: "the true lie." "The lie in words," like deceiving an enemy or inventing a fable, may be useful. "The true lie" designates an "ignorance about the highest realities" in the soul of him who believes sincerely that he is acting rightly. "The true lie is not useful; it is hateful." Not knowing any better excuses nothing, even though it may explain much. Plato grants no standing to the sincerity plea, as behooves a philosopher who often attributes all virtues to knowledge.

and probing conversations to point up and offset the monster Raskolnikov carries within him. Unlike Camus, Dostoyevsky does not indict the judges and the jury and does not create undue sympathy for the criminal. Raskolnikov is convicted and serves his sentence at hard labor. Through this penance and through his faith in Sonia, he may find forgiveness and redemption. Because we see him in the round, we understand Raskolnikov far better than we do Meursault.

The comparison of *Billy Budd* and *The Stranger* affords a rare opportunity for literary and psychological diagnosis. The former novel helps correct the common misreading of the latter. We see the truth most clearly when it comes to us in the form of an error dispelled. *The Stranger* enacts a scandalous moral irony, though difficult to detect in the narrative of a near-autistic man. Meursault, the modest white-collar worker for whom most modern readers come to feel strong empathy, really is a monster. Camus has written the equivalent of a moral labyrinth, from which some readers will not escape.

"A sin," writes Coleridge in *Aids to Reflection*, "is an evil which has its ground or origin in the agent, and not in the compulsion of circumstances" (Aphorism X, Comment). Billy and Meursault lack a strong sense of self and of agency. Therefore, the murders they commit can be painted to look like the consequence of "the compulsion of circumstances." Camus, by composing an "inside narrative" in the first person, links our sympathies so closely with the tiny sensuous rewards and the general affectlessness of Meursault's life that we may lose our moral bearings. That is why the proverb "To understand is to forgive" can strike us—wrongly—as indulgent rather than cautionary.

5. Knowledge as Interference

But how much can we understand? Can we ever peer beyond the caul of selfhood that enfolds us by the time we begin to talk and to answer to our name? Can we know another person? Our tentative answers to those questions usually fall into a few areas of inquiry known as literature and philosophy and history, areas we lump together as "the humanities." Across the centuries, the humanities have offered a broad set of answers fluctuating between faith and doubt. Ours has been predominantly a century of doubt. One of our most philosophical novelists, Marcel Proust, answers those fundamental questions in the negative by borrowing a metaphor from physics.

> *When I saw an external object, my consciousness that I was seeing it remained between me and it, outlining it with a narrow mental border that prevented me from ever touching its substance directly; in some way the object volatized before I could make contact, just as an incandescent body approaching something moist never reaches moisture because of the zone of evaporation that always precedes such a body.*
>
> (I, 84)

In this description our isolation is inescapable. But a few pages later, Proust made an exception for works of literature: Their transparency permits us miraculous entry into other lives composed not of opaque flesh but of comprehensible words. Thus we can pose the fundamental questions again and with different results concerning the two literary works *Billy Budd* and *The Stranger*. Their carefully composed sentences manipulate our attention, understanding, and sympathy in very specific ways. And now we must acknowledge, as I have suggested earlier, two markedly different senses of the infinitives *to know* or *to understand*.

Though Billy is given stronger status as the legendary Handsome Sailor than as a realistic particularized individual, the essentially third-person narrative approaches very close to his consciousness. We feel the menace behind the triviality of the soup-spilling scene; we share Billy's impatience with the after guardsman's attempt to talk to him alone in the lee fore chains. But by the time we read the central scene of Claggart accusing Billy in the captain's presence of conspiring to mutiny, we have enough familiarity with Billy's temperament and vocal handicap to understand his striking Claggart a blow that turns out to be fatal—a blow in the name of simple truth. Captain Vere's first response—"Fated boy," he utters—informs us both that he understands Billy's explosive outburst well enough to forgive it and that nevertheless Billy will have to suffer the full legal consequences of his insubordination and homicide. Melville adjusts the story line so as to carry us just far enough inside Billy and Vere to allow us to deplore and to accept the tragic outcome. Claggart remains a "mesmeric" mystery. Our understanding of Billy and of Vere does not, in an intelligent reading, paralyze our judgment. Our understanding complicates and enlarges our judgment.

The Stranger has the opposite effect on most of its readers. By the time we reach the central scene of the shooting on the beach, our point of view has adapted itself to Meursault's passiveness before other people's initiatives and before the sheer momentum of an episode once started. We probably accept the metaphor that the whole landscape heaves up ("*C'est alors que tout a vacillé*") and propels him toward the fatal act. We are swayed and blinded by the circumstantial narrative to the point of overlooking the pit of monstrosity that opens up around his action. The students in their papers and Camus in his preface have "understood" in the sense of empathizing with Meursault's absorption in a pure present without history or responsibility. That perspective impedes their judgment even during the second part, when Meursault gains enough detachment to glimpse his own guilt.

In its impact on people's behavior and sense of "alienation" and by its apparent sincerity of feeling, *The Stranger* came close to becoming the mid-twentieth-century equivalent of Goethe's bestselling *The Sorrows of Young Werther* (1774), which provoked hundreds of suicides all over Europe. Werther cannot explain to himself the sentiment he has of disintegration and decline from the ideal nobility of heart inspired in him by Lotte's perfection. His rebellion turns him first against the society that supports him and then against himself in a carefully staged suicide. *Werther* and *The Stranger* are excessively romantic and self-absorbed stories verging on solipsism. Meursault has no inkling of how estranged he is from human life until he is arrested for destroying another life. Then his brief rebellion burns itself out on a well-intentioned priest, and Meursault tries to transform his execution into a symbolic suicide by choosing it, by welcoming it. Werther's overreaching of sentiment in the eighteenth century collapses into

Meursault's twentieth-century listless "absurd," and the influence of the latter echoes the enormous vogue of the former. If Camus had told us the color of Meursault's shirt, it would have set off a fashion as widespread as that of Werther's yellow waistcoat.

Camus perfected a hypnotic prose style combining Hemingway's laconism, Kafka's sense of tragicomic inscrutability in all things, and Voltaire's deadpan portrayal of naïveté in *Candide.* The resulting novella entraps us all on first reading. We feel more fascination than horror at the course Meursault's life takes. Our inside knowledge of his deeds becomes a form of bewitchment or possession, difficult to exorcise, leading us to suspend judgment, even to unseat justice. A small leap of association permits us to read beyond the highly charged literal meaning of *The Stranger* to discern a double parable about contemporary events that were taking place in the Soviet Union and Nazi Germany. Meursault stands for the citizen whose passiveness and stunted imagination allow him to yield to outside pressures to carry out inhuman action. And readers who sympathize unthinkingly with Meursault stand for potential accomplices and collaborators in his actions. The power of the book's political significance arises from the fact that it remains entirely implicit. The risk of misreading Camus' novel lies in the appeal of the empathetic knowledge it offers us of an enigmatic character only too easy to identify with. Is there, consequently, a point at which we must beware of such knowledge? Beware of empathy?

The two novels I have been discussing, particularly *The Stranger*, with its suggestive title, lead us to one of the more distressing categories of forbidden knowledge. The closer one approaches to an event or to a person, the less securely one seems to know it. The trees obscure the forest. The more one knows, the less one knows. Perception itself requires a certain distance. Empathy hides more than it reveals.

More than most modern philosophers, Isaiah Berlin strives to reconcile empathy for others with reasonable standards of decency and moral behavior. But when in "Historical Inevitability" he carefully glosses the proverb "To understand is to forgive," Berlin has the honesty to write, "To understand all is to see that nothing could be otherwise than as it is." In other words, to understand all is to capitulate to the status quo. That proposition precisely describes the character of Pangloss in *Candide.* What Berlin presents discursively in his essays, Melville and Camus approach very differently in their fiction. In both kinds of writing, the author is thinking in terms of imaginary situations, is conducting thought experiments in order to reflect upon the meaning and worth of human actions. The difference is that whereas Berlin carries out the crucial parts of his thought experiments in terms of freedom, authority, natural rights, and other abstractions removed from persons and from time, Melville and Camus create their thought experiments on empathy (particularly on the reader's empathy) in terms of fully conceived characters subject to all the contingencies of time and mortality. They stage an action to show how knowing too much can affect us, perhaps blind us, even when we have gained that knowledge from the essential faculty of empathy. The double bind of empathy as laid before us by *Billy Budd* and *The Stranger* points us finally in the direction of a middle way between moral certainty and moral ignorance.

In the opening pages of *The Myth of Sisyphus*, Camus speaks of seeking to know people by analysis of their actions, both sincere and insincere. Then he concedes, "The method defined here admits to the feeling that any true knowledge is impossible." But Camus exaggerates. He and we know the truth that Meursault murders a man and deserves punishment. But we shall never know exactly *why* he did it. In trying to fathom that mystery of iniquity, we can lose our way and come to find no fault and no guilt in so sincere and so unassuming a temperament. Melville's novel helps us to see in Camus' tale the severe interference between two types of knowledge: inside and outside. Between them, the two stories offer a striking moral education.

INTERLUDE: TAKING STOCK

■

1. Forbidden Knowledge and Open Knowledge

Every story I have discussed deals directly or indirectly with the human trait of curiosity. Curiosity, in turn, leads almost fatefully to the theme of forbidden knowledge, to a potential limit on curiosity. Curiosity drives particular individuals to actions like those of Pandora and Psyche, of Dante's Pilgrim, and of the Elephant's Child. In *Paradise Lost,* Eve's dreamy imagination turns curiosity into a form of lyric subversiveness. Out of fear of losing everything, the Princesse de Clèves restrains her yearning to discover further dimensions of love. Faust's ambition carries curiosity into the realm of what the Greeks called *pleonexia*—insatiably wanting more than one's due. With Meursault, the *lack* of curiosity about himself or anyone else produces the illusion of "sincerity," which masks the inhumanity of his behavior. I discern no clear progression in these stories. We see, rather, the fluctuations of a dominant motive.

Now curiosity carries within itself a principle of doubt—doubts about received knowledge and the conventions of the status quo. After Galileo and Descartes, the principle of doubt has spared nothing, not even curiosity itself. Thus in certain illuminating situations, curiosity has had to acknowledge its own limits. "Be lowly wise" emerges as the moral of Milton's version of Adam and Eve's story. Pascal recommends that we "know our reach *[portée]*." Out of loyalty to family and friends, Maggie resists her two suitors in *The Mill on the Floss*. Huxley coined the word *agnostic* to designate a limit on both his scientific and his religious beliefs. All these episodes stand for some form of limit imposed on the "wayward" faculty of curiosity. In that respect, the stories I have been discussing contain the theme of forbidden knowledge as a common denominator.

Something disturbing happens, however, when any limit is imposed from outside or by apparently arbitrary prohibition. I have called this response the Wife of Bath effect: "Forbid us thing, and that desire we." Such an impulse restarts the whole cycle of curiosity by provoking a newly defiant *libido sciendi*. Why this perverseness? Why this skittishness before any restriction imposed not only on actions but even on knowledge?

I have suggested two answers, similar but not identical. *Billy Budd* twice uses the biblical phrase "the mystery of iniquity." It refers to a strain of thwarted humanity in both Billy and Claggart that drives them to an action ultimately harmful to themselves. But both they and we remain partly ignorant of the existence and nature of iniquity in their character. Melville does not offer a solution to that mystery.

Nicholas Rescher's phrase "the fog of uncertainty" suggests another reason for the Wife of Bath effect. One of the basic givens of humanity is final ignorance about ourselves and those closest to us. But we cannot help kicking against this aspect of the human condition, wanting to know what we can never know. Consequently, out of impatience and sheer human orneriness, we yield to the Wife of Bath effect. In the concluding chapter, I shall return to this subject in discussing "the veil of ignorance."

For social and religious reasons, earlier ages accepted more readily than we do some form of forbidden knowledge. Most thinkers, though not all, made their peace with restrictive notions like taboo, the *Index*, heresy, *arcana dei*, and Bacon's "proud knowledge." Today, we describe two periods of history as having loosened and even overthrown such constraints. In our version of the past, the Renaissance and then the Enlightenment introduced an opposing ideal of *open knowledge.* Early in the Renaissance, Pico della Mirandola appeared almost to foresee the future all the way to evolutionary theory. Pico described man as having "no fixed seat," as being "the molder and maker of thyself . . . who canst again grow upward from thy soul's reason into the higher natures which are divine." Pico's visionary humanism prepares the way for Descartes' affirmation of the principle of doubt, not faith, as the starting point of reason. When these heretical doctrines combined with the gradual secularization of life, plus the printing press and the beginnings of free speech, then a major countervailing force had taken shape to oppose the notion of secrets—secrets of God or of nature. Open knowledge appears to stand for modernity itself. Kant borrowed an injunction from Horace to begin "What Is Enlightenment?": "Dare to know!"

Open knowledge as a modern achievement appears to have left behind the tradition of esoteric knowledge only for initiates.* Today, the principle of open knowledge and the free circulation of all goods and ideas have established themselves so firmly in the West that any reservations on that score are usually seen as politically and intellectually reactionary. However, the stories examined in the preceding chapters demonstrate in diverse ways that the principle of open knowledge has not everywhere driven out the principle of forbidden knowledge.

2. The Consequences of Open Knowledge

What are the unforeseen consequences to society and to individuals of forbidden knowledge thrown open? A foreboding tale I discussed in the Foreword dramatizes a nineteenth-century answer to the question. The number of recent theatrical, film, and TV adaptations of *The Strange Case of Dr. Jekyll and Mr. Hyde* testifies to the

*Both the fear of persecution and an exclusivist sense of the truth as sacred, as forbidden fruit, led earlier authors to "write between the lines," as Leo Strauss describes it. And even an Enlightenment figure like Lessing was still "concerned that there are truths that cannot or should not be pronounced." Appendix II is devoted to a discussion of forbidden knowledge and the occult.

continuing appeal of Stevenson's story. The gruesome episodes, the self-absorption of the hero, the portentous style, the gothic setting of scientific investigation bordering on magic—all these elements alert the reader to the connections of the Jekyll-Hyde parable to the Faust-Frankenstein pair and to fictional treatments of the double by Hoffmann, Poe, and Wilde.

But there is a stronger reason to call attention to *The Strange Case of Dr. Jekyll and Mr. Hyde.* Its strangeness and its familiarity arise not only from the haunting theme of the double hidden in each of us but also from the way the events of the story project convincingly around them two significant motifs: scientific experiment gone awry and prurient interest. The first lies fully visible on the surface and supplies, in the form of Dr. Jekyll's experimental drugs, the presumed explanation for the mysterious events that ruin a reputation and divide a single individual against himself. The second motif is kept muffled by Stevenson behind the phrase "undignified pleasures." Only in the closing pages do the words *monstrous, bestial,* and *torture* imply the dimensions of evil that Jekyll can vicariously indulge in through his proxy, Hyde. Readers have not neglected these clues. All later remakings of the story in many forms have more or less explicitly identified Hyde with criminal violence associated with sexual excess—usually by adding a victimized woman not present in Stevenson's pages. We should not overlook the fact that Stevenson wrote the work while the Darwinian controversy was raging over the beast within us—whether such animal proclivities represent, as in the Christian tradition, a test of our spiritual will to control them, or whether they represent a worthy portion of our compound nature too sternly reduced in the past by the constraints of religion and traditional morality.

There is more than sheer coincidence in the fact that *The Strange Case of Dr. Jekyll and Mr. Hyde* appeared in 1886, the year of Nietzsche's *Beyond Good and Evil.* When he puts aside "all sentimental weakness," Nietzsche declares directly what Dr. Jekyll can express only by delegating deeds to Mr. Hyde. In Nietzsche's view, one may occasionally display consideration for others purely as good manners among privileged peers. But he endorses a quite different standard of general moral conduct: ". . . life is *essentially* appropriation, injury, overpowering of what is alien and weaker; suppression, hardness, imposition of one's own forms . . . exploitation" (§259). Nietzsche celebrates the triumph of Hyde over Jekyll.

In the next section, I pursue the two further aspects of my subject suggested by *Dr. Jekyll and Mr. Hyde*: scientific experiment gone awry and prurient interest. Chapter VI considers a number of incidents in the history of science and the challenge they throw down to any constraints implied by the notion of forbidden knowledge. Chapter VII approaches the vast problem of evil lodged in our nature and, if released, forever prepared to regain control of our actions. I do so by examining the life and writings of the Marquis de Sade and his rehabilitation in the twentieth century.

Both chapters engage in highly contentious issues—mostly moral.

PART TWO

CASE HISTORIES

CHAPTER VI

KNOWLEDGE EXPLODING: SCIENCE AND TECHNOLOGY

■

1. THE BOMB AND THE GENOME

"*In some sort of crude sense . . . the physicists have known sin.*"

—J. ROBERT OPPENHEIMER, 1947

"[The Human Genome Project] *is the grail of human genetics . . . the ultimate answer to the commandment, 'Know thyself.'* "

—WALTER GILBERT, 1986

Dramatic enough as separate statements, my two epigraphs set side by side form a shocking contrast. Does scientific research, backed by immense technological and political support, represent the ultimate *sin* of Western civilization? Or is it the *grail* we seek as our only remaining form of salvation? Do these two statements express legitimate interpretations of the present status of science and technology?

After five chapters devoted primarily to exploring forbidden knowledge in legends and stories, it is time to deal with its manifestation in a set of events that press more and more directly on our daily lives. In the following pages on science, I do not follow chronological order. Modern experimental science goes back only to the seventeenth century. In that short span, its influence has grown so rapidly that it appears to rival the influence of religion and the state. Science comes at us from all directions.

Albert Einstein's letter in 1939 to President Roosevelt on the need to construct an atomic bomb was written in the face of an unprecedented attack on civilization from within our combined

Judeo-Christian, Greco-Roman heritage. Einstein's reluctant yet urgent call to action (prompted by Leo Szilard) led to a collaboration of science, technology, and entrepreneurship that bears comparison with the construction of the Pyramids, the Great Wall of China, and the Panama Canal. More concentrated than those undertakings because of the urgency and the secrecy of its mission, the Manhattan Project was carried out within stark Aristotelian unities of time, action, and character. That drama presents us with a tragic hero who succeeded brilliantly in his assigned task, yet who ultimately saw himself as having betrayed the trust of his high calling—a frail, fedora-wearing Prometheus, a chastened Frankenstein.

J. Robert Oppenheimer was a respected theoretical physicist, an organizer-director who earned the loyalty of hundreds of scientists, technicians, and military personnel, and a troubled philosopher of the responsibilities of his mission.* In retrospect, Oppenheimer appears to have been typecast to express the rival convictions that grew up within the project. The work of those involved would help defend democracy and human decency against a new form of state barbarism. Their work would also unleash a force so destructive that only fears that the enemy might discover it first justified the effort. We know the crucial roles played by Einstein and Fermi and other scientists, by Roosevelt, General Groves, and Truman. But in our minds, we have projected upon Oppenheimer the responsibility to answer two distinct questions: "Shall we manufacture the bomb? Shall we employ the bomb?" Oppenheimer became both

---

*When the first test bomb exploded at Alamogordo, New Mexico, Oppenheimer had chosen and learned lines so appropriate to his part and to the historic moment that many students now learn them in school. Both quotations come from the *Bhagavad-Gita*.

For the flash:

If the radiance of a thousand suns
were to burst into the sky,
that would be like
the splendor of the Mighty One.

And for the mushroom cloud:

Now I am become Death, the destroyer of worlds.

hero and scapegoat for having answered the two questions in the affirmative. In June 1945, he rejected the scientists' Franck Report, which opposed any unannounced use of the bomb. After Hiroshima, he changed his mind. He is our Hamlet. Later public questionings of his loyalty and denial of his security clearance only enhance the portrait of a person racked by the disputes of his time.

Just two years after Hiroshima and Nagasaki, Oppenheimer was invited to deliver a lecture at the Massachusetts Institute of Technology. He gave it the neutral title "Physics in the Contemporary World." Everyone was highly conscious in 1947 of Oppenheimer's appearing as the ex-director of the Manhattan Project before an audience of scientists at a historic moment as the world emerged from World War II. He had composed a subtle and highly personal manifesto in defense of science. After affirming that "physics is booming" especially in the field of elementary particle research, he applied the elusive principle of complementarity to science itself. In other words, he described two conflicting interpretations and affirmed the truth of both. They complement each other as partial, not exhaustive, truths. On the one hand, the value of science lies in its fruits, in its effects, more good than bad, on our lives. On the other hand, the value of science lies in its robust way of life dedicated to truth, disinterested discovery, and experiment. The practicing scientist feels a greater kinship with the second principle; he is at best "ineffective" when he tries "to assume responsibility for the fruits of his work." That task is properly assumed, Oppenheimer declared, by statesmen and political leaders. One wonders if he had read *Frankenstein* along with the *Bhagavad-Gita*.

In this context of affirming scientific research, the most widely quoted passage in the talk seems surprisingly out of place, as if Oppenheimer could not bring himself to exclude it from an otherwise-affirmative statement of the strengths of the scientific approach to knowledge. The passage must have jarred his listeners in 1947 even more than it jars a reader today. After mentioning "a legacy of concern" left by World War II and the development of the atomic bomb, he inserted this alien paragraph.

Despite the vision and the far-seeing wisdom of our war-time heads of state, the physicists felt a peculiarly intimate responsibility for suggesting, for supporting, and in the end, in large measure, for achieving the realization of atomic weapons. Nor can we forget that these weapons, as they were in fact used, dramatized so mercilessly the inhumanity and evil of modern war. In some sort of crude sense which no vulgarity, no humor, no overstatement can quite extinguish, the physicists have known sin; and this is a knowledge which they cannot lose.

(GARDNER, 193)

At the center of Oppenheimer's encouragement to young physicists squats an ominous monster of guilt. He refused with impatience any distortion or dismissal of it by nervous joking or excessive breast-beating. This nonreligious scientist could not have found a stronger word than *sin* to express a conviction about complicity with evil. He had not opposed the policy decisions that led to immense destruction of civilians at Hiroshima and at Nagasaki. The "knowledge" referred to in the last sentence is of a different order from that of scientific knowledge. Oppenheimer meant *moral* knowledge. He appeared to be on the verge of propounding a Hippocratic oath for scientists.

Having confessed his guilt and acknowledged the consequences of his deeds, Oppenheimer returned to the generally optimistic message about science, even though he granted that science could not establish a secure peace. This embedded cautionary passage in his MIT lecture corresponds to a sentence Oppenheimer blurted out during a friendly conversation with Truman after the war. "Mr. President, I have blood on my hands." Truman was exasperated by what must have appeared to him as Shakespearean posturing.

At the United Nations in June 1946, the Soviets vetoed the Baruch Plan for banning atomic weapons, destroying United States atomic bombs, and vetoless international control of all atomic materials. By the time the Soviets exploded their own atomic bomb in 1949, the Atomic Energy Commission, the Rand Corporation (contracting weapons research and development for the new Air Force), and the indefatigable physicist Edward Teller were working at a new order of magnitude. The destructive force of the hydrogen bomb detonated in 1952 on the atoll of Eniwetok had to be measured not in kilotons but in megatons of dynamite—one thousand times more powerful than the Hiroshima bomb. Oppenheimer's opposition to the development of the hydrogen bomb earned him classification as a "security risk" and dismissal from his responsibilities advising the Atomic Energy Commission. Teller knew none of Oppenheimer's scruples about the possibility of sinful knowledge. In a 1994 interview, Teller, looking back at the H-bomb debate, pronounced a sentence in which it would be difficult to distinguish optimism from *pleonexia*. "There is no case where ignorance should be preferred to knowledge—*especially* if the knowledge is terrible."

In the 1990s, without the Nazis or the Soviets to propel us, we appear eager to reverse the proliferation and development of nuclear weapons. But the danger of future development and use by other countries has probably increased. Meanwhile, we have embraced another massive collaboration involving scientific research, technology, and government. This time, the project is peaceful and includes a major role for commercial and corporate interests as propelling forces. Four decades of genetic research on DNA have brought us to the point where journalists and scientists speak the same optative language about cracking the code of life, reading the human blueprint, and filling in the map of human nature. The ambitious Human Genome Project, voted and funded by the U. S. Congress in 1988, will cost considerably less than the budget for Project Apollo and probably will furnish more sheer data than the moon landing for scientists to reflect upon.

However, the project has also provoked severe criticism. The promised medical rewards to individuals in terms of therapy remain uncertain, partially because concentration on mapping and sequencing does not solve the challenge of gene replacement when an undesirable gene is found. Meanwhile, prenatal and carrier testing permitted by newly developed technologies will complicate our lives in ways that society is little prepared to cope with. As increasing numbers of fetuses are diagnosed with serious disorders, abortion has become a widely practiced therapeutic procedure, an elective, ad hoc version of sterilization, which was formerly favored by eugenicists. Claims for and against the HGP reached peak intensity around 1992, and judicious criticism succeeded in preventing it from becoming a crash program to solve a limited problem.

Looking back, we can examine the way concerned parties in the Manhattan Project and the Human Genome Project spoke about their undertaking and its importance to the society. Required to operate in total secrecy, the Manhattan Project justified itself as the last resort of a civilization driven to the wall by malevolent forces. Nevertheless, esprit de corps and a clear sense of purpose in opposing totalitarianism could not extinguish the mutterings of conscience among key participants over the human and political consequences of the new weapon. Oppenheimer's recourse to the Christian notion of sin to characterize those doubts was melodramatic but not without justification.

With no international crisis to sustain them, the proponents of the HGP have generally chosen a hyperbolic mode of expression. The quotation below appropriates not only a popular Christian symbol but also a celebrated Socratic aphorism. The image of a beleaguered defender of civilized values reluctantly resorting to a destructive weapon is now displaced by that of a prophet offering a panacea for all our ills. "[The Human Genome Project] is the grail of human genetics . . . the ultimate answer to the commandment 'Know thyself.'" This is a quote from Nobel laureate biologist Walter Gilbert, who spoke at the 1986 Santa Fe meeting organized by Charles DeLisi of the Department of Energy to consider sequencing all our 3 million base pairs. In later years, Gilbert did not mute his triumphalist rhetoric.

I propose now to set Oppenheimer's and Gilbert's statements in perspective and to consider how science and technology in the twentieth century have both recognized and flouted the notion of forbidden knowledge in its several forms. I first discuss one preliminary matter I have already raised and then take up several instances of limits applied—and not applied—to scientific undertakings.

2. The Siren Song: Pure and Applied Science

Oppenheimer's differentiation in his 1947 lecture between science as a disinterested pursuit of truth and science as an activity having profound effects on our lives restates an almost universally accepted distinction with a long history. It draws on a parallel distinction that took shape in the Catholic church between the monastic calling and the pastoral calling for priests. Francis Bacon, the seventeenth-century apologist for scientific inquiry, wrote a short fable on the subject, and it turns into a cautionary tale. "Sphinx, says the story, was a monster," Bacon begins. After briefly retelling the Sphinx story, he jumps headlong into interpretation.

> *The fable* [of Sphinx] *is an elegant and a wise one, invented apparently in allusion to Science; especially in its application to practical life. Science, being the wonder of the ignorant and unskilful, may be not absurdly called a monster. . . . Again Sphinx proposes to men a variety of hard questions and riddles which she received from the Muses. . . . when they pass from the Muses to Sphinx, that is* from contemplation to practice [emphasis added], *whereby there is necessity for present action, choice, and decision, then they begin to be painful and cruel . . .* *

If the Sphinx was a monster, then science is a monster also—presumably of our own creation. We do not expect such a view from Bacon. But he explains himself clearly. As the Sphinx grafts a human head onto a lion's body and combines the contemplative questions of the Muses with "painful and cruel" choices posed by its own riddles, so science must pass "from contemplation to practice." Here, Bacon is *not affirming* the conventional distinction between pure research and its applications, a distinction that has become the principal defense of science against the challenge of forbidden knowledge. In the figure of the monster, he is acknowledging the intention to distinguish separate parts and is showing that the separation is impossible in practice. Bacon wants us to know that the Sphinx exists and represents the menacing side of science. Though they form an unnatural combination of parts, the dreaming human head cannot be separated from the lion's menacing body. That graft constitutes precisely the monstrousness of the creature, of which we should beware. For the imaginings of the mind will inevitably be given "present action" by the attached body: a monstrous union, symbol of the bond between science and technology.

We go to great lengths to deny Bacon's warning. A phalanx of official societies and institutions and universities devote themselves

*Appendix III reproduces the entire story.

to basic research. Technology and engineering are defined as applications of a purer and more basic knowledge. In an issue of *Daedalus* devoted to "The Limits of Science" (Spring 1978) the essay by the molecular biologist David Baltimore argues strongly against any limits on pure science. He does not agree with Bacon or acknowledge any difficulties in severing the head of the Sphinx from its body. "I want to make a crucial distinction. The arguments [for unrestrained freedom] pertain to basic scientific research, not to the technological applications of science. As we go from the fundamental to the applied, my arguments fall away." But shall we ever be able to draw a firm line—or even a rough line—between discovery and application? If we can, should we argue that responsible scientists will remain on the pure side of the line? That was Oppenheimer's argument in his 1947 lecture. Presumably, other agencies and institutions would decide about whether, when, and how to use the discoveries. But what might at first appear to be a reasonable position does not hold up for long. At the end of the twentieth century, few problems grip us more importunately than this one.

At what point could work on the atomic bomb have stopped in order to win the war without any unnecessary loss of human life? Among those in the know at Los Alamos, the debate raged with intensity and in great secrecy. Should international observers have been invited to Alamogordo? We feared a dud. Should we have dropped the first airborne bomb on an uninhabited island near Japan? We had only two bombs and could not "waste" one. We did not know where or how to separate the head from the body. Today we face the immense and very different problems of how to "apply" the "pure" knowledge gained from molecular genetics and the Human Genome Project. But research scientists themselves have already obliterated the line of distinction by participating in commercial enterprises to exploit markets for genetic knowledge. Baltimore and many like him speak as if we can observe in scientific research a principle of separation like the one that regulates the three branches of our government and that divides church from state. But we should heed Bacon's caution about believing we can easily separate contemplative or pure thought from its application to our lives. Our current scientific institutions do not succeed in doing so. The history and theory of patent law reveal that its essential function is to encourage the wide commercial exploitation of salable discoveries rather than to restrict and protect new knowledge. Individual investigators, from Galileo and Leonardo to Oppenheimer and James Watson, have not observed the distinction in their own lives and work. Even Einstein, the ultimate figure of pure research, whose laboratory was a blackboard, felt compelled to write a letter to the President of the United States urging the application of atomic theory to constructing a bomb. The frontier between pure and applied is a phantom that appears on many maps yet cannot be located easily on the terrain.

There is an ancient epic, one of whose episodes describes what could be interpreted as an elaborate strategy to obtain dangerous knowledge without the ability to act on it or to apply it in any way. In Book Twelve of The *Odyssey*, Odysseus is warned by Circe about the song of the Sirens, which he will soon encounter. If he and his crew yield to the bewitchment of that music, they will perish. Circe instructs him how, if he wishes to listen, to do so without succumbing. Why doesn't she tell him to plug his own ears with wax, as he instructs his faithful sailors to do? Why does Odysseus take up the possibility of hearing this mortally perilous song? Why does Circe accept and possibly admire his privileged curiosity about something he does not need to know? Finally, why in helping him does she pander to his curiosity when she knows that, by himself, he lacks the strength of will to resist the Sirens' song?

The *Odyssey* presents an ambiguous universe inhabited by gods, demigods, and those favored by the gods. Circe's instructions leave Odysseus' perceptions and mind free while restraining his body from response. She grants him the possibility of knowledge protected by a safe distance, without the immediacy of direct exposure. That would cost him his life. In advance and after the fact, when safe from actual temptation, Odysseus is content to accept this indirect or incomplete knowledge. Homer narrates the episode to his reader or listener with a cautionary gesture. "Listen with care to this," he quotes Lady Circe as saying to Odysseus, "and a god will arm your mind." When Odysseus later recounts the encounter to Alkinoos and his court, he begins: "More than one man, or two, Dear friends, should know those things Circe foresaw for us."

The strategy of committing oneself irretrievably in advance to a restricted line of conduct serves as an object lesson for all weak-willed persons. By quoting six stanzas of the Sirens' song in his narrative, Odysseus demonstrates his successful conversion of lethal knowledge into safe knowledge. He achieves it through self-imposed constraints according to privileged advice.

Odysseus' survival relies on a carefully arranged boundary drawn between knowledge and what in earlier chapters I have called "experience."* But the situation cannot last. He develops no permanent immunity to the temptations of mortal life. The Sirens' episode draws a boundary between pure and applied knowledge, but only with divine assistance to establish tight interlocking constraints on the operations of inquiry. Without those conditions, most intellectual inquiry takes place on a slippery slope between pure knowledge and the likelihood of its application in the real world.

The conception, development, and use of the atomic bomb offers the starkest illustration. We now know that Roosevelt's 1939 decision to build the bomb was to have been followed, if the project was successful, by "mature deliberation" about its possible use. But six years later, under a new President, as Alan Cranston has pointed out, "the nuclear chain of events had assumed a life of its own." We were already carrying out horrible incendiary bombings of Germany and Japan. Many strategic considerations impelled us forward. Truman did not interfere with the ongoing military planning for dropping this new bomb on Japan. In other words, once the process of development was started, the military use of the bomb virtually decided itself by a kind of technological momentum. Mature deliberation came not before but after the fact. From Einstein's talismanic $E = mc^2$ to Hiroshima turned out to be an increasingly slippery slope little affected by good intentions and individual twinges of conscience.

We have no divinities to intercede for us today. Who, if anyone,

---

*It would be far from wrong to conclude that Odysseus wanted to have it both ways. For, as he travels past the Sirens lashed to the mast, he grants himself a form of knowledge without responsibility. In thus abstaining from full experience, he follows a course of action not unlike that of the Princesse de Clèves and of Emily Dickinson's persona as presented in Chapter IV. All three figures advance a long way toward the ultimate ordeal of life—and stop short of the final authenticating or destructive step. By withdrawing, they hope to preserve freedom of imagination from too particular and binding applications. They achieve a certain haughty integrity.

Jon Elster has many worthwhile observations to make about the Sirens and the social sciences.

could or should bind our scientists to a mast? David Baltimore closes the paragraph from which I quoted earlier by stating: "There are many technological possibilities that ought to be restrained." In the same article, he accepts modified constraints even on pure research. "Society, while it must determine the pace of basic scientific innovation, should not attempt to prescribe its directions." Without a goddess to guide us, however, the line between pure and applied research and the line between too fast and too slow in opening up a field become very difficult to establish. We tend to believe that any scientific truth worth knowing has significance for our lives and that to restrict either its pursuit or its application contradicts the very nature of truth.

The wisest place to turn to illuminate these dilemmas is not to speculation but to specific case histories.

3. LIMITS ON SCIENTIFIC INQUIRY: FIVE CASES

The question of possible limits on science is, of course, a modern version of the constraints resisted by both Galileo and Bacon at the beginning of the seventeenth century. On the surface, Galileo acquiesced to the Church's demand that he recant his belief in Copernican heliocentrism and cease this line of investigation. Nevertheless, under house arrest, he continued his work and contrived to smuggle out his writings to Holland. Bacon's compromise was more subtle. As I have described in the first chapter, *The Great Instauration* affirms the virtue of "pure" knowledge learned from the divine book of nature and renounces "proud" knowledge that would trespass upon theology and revealed religion. Between them, Galileo and Bacon set modern science on its path of independence from religion and, to a lesser degree, from philosophy.

Three hundred years later, in November 1945, just three months after the United States dropped two atomic bombs on Japan to end World War II, the Association of Los Alamos Scientists met to hear a talk by Oppenheimer, their director. Many of them were horrified by the way the bomb had been employed. All of them were asking themselves, What have we done? Some were thinking about the need to prevent science from causing further destruction on such a scale. In the paragraph that responds most directly to these humanitarian and professional scruples, Oppenheimer affirmed the autonomy of science even more firmly than Galileo and Bacon.

But when you come right down to it the reason that we did this job is because it was an organic necessity. If you are a scientist you cannot stop such a thing. If you are a scientist you believe that it is good to find out how the world works; that it is good to find out what the realities are; that it is good to turn over to mankind at large the greatest possible power to control the world and to deal with it according to its lights and values.

(RHODES, 761)

Oppenheimer was also being as devious as Galileo and Bacon. The bomb project was a technological application of basic science, not a new discovery. His "organic necessity" comes very close to being the patriotism of scientists. It may be "good to find out how the world works," but it is by no means evident that the "lights and values" of mankind at large would have ratified the Manhattan Project's transforming that knowledge into an atomic bomb to be dropped on two populous cities.

Toward the end of Chapter I, I referred to the writings of a contemporary philosopher, Nicholas Rescher, who wrote a book entitled *The Limits of Science* (1984).* In another essay—given the blunt title "Forbidden Knowledge"—on this vexed question, Rescher balances arguments for regulating science as a form of power against arguments for adopting a completely laissez-faire policy. One of his fundamental propositions, which could apply to pornography as well as to science, appears to clear the air immediately. I quote his words again.

There seems to be no knowledge whose possession is morally inappropriate per se. *Here inappropriateness lies only in the mode of*

*J. W. N. Sullivan's *The Limitations of Science* (1933) presents, in spite of its title, an account of the promises and values of science. Useful surveys of the antiscience tradition can be found in John Passmore, *Science and Its Critics* (1978) and in Gerald Holton, *Science and Anti-Science* (1993).

acquisition or in the prospect of misuse. With information, possession in and of itself—independently of the matter of its acquisition and utilization—cannot involve moral impropriety.

(FORBIDDEN KNOWLEDGE, 9)

Rescher goes on to acknowledge that the pursuit and application of scientific knowledge can at times cause disadvantages and dangers to the common good. Following a middle-of-the-road position, he suggests a set of categories that I have modified and expanded into five considerations that could justify imposing constraints on scientific inquiry. I arrange the five categories in the order of increasing complexity and illustrate them from different periods in the history of science. Later, I shall return to the question whether *any* knowledge, in itself, can be immoral or evil.

PRACTICAL CONSIDERATIONS: ARCHIMEDES

In the third century B.C., the Greek physicist and mathematician Archimedes earned his niche in history, so the story goes, by shouting "Eureka" in the bathtub. He had just figured out how to tell a pure gold crown from alloy by its specific gravity. Contemporary historians also record that Archimedes boasted to King Hieron II of his native Syracuse, "Give me a place to stand, and I will move the earth." For a moment, let us treat this sentence not only as unbeatable advertising copy for the principle of the lever but also as the earliest recorded grant application in physics. Hieron's peer-review panel would have had a whole set of reasons to turn Archimedes down. He had not specified any workable position for himself or his fulcrum. He hadn't even mentioned the required size of the lever and how it would be held in place and moved. How long would the experiment take? How much would it cost? Hieron had no resources at that point to commit to a project of dubious significance and no evident benefits to the city of Syracuse. Furthermore, members of the panel might have suggested that Archimedes had made his point already just by imagining the experiment. Funding and performing it would amount to a vast boondoggle, robbing other worthy claimants on the public treasury.

For practical reasons, then, Archimedes would not have been funded if he had asked for a grant. Two thousand years later, in comparable circumstances, Einstein did not request funds to set up real elevators in space in order to demonstrate the principle of general relativity. Like Archimedes, he knew that a "thought experiment" could in some circumstances serve the purpose equally well. In our time, expanded national budgets, international competition, and the enormous complexity of scientific projects have changed the scale on which we envision scientific research. Advocates of space research and of a supercollider have presented their projects as virtually inevitable. But practicality and priority are still major factors even for a modern Archimedes with a dramatic scheme to move the world.

PRUDENTIAL CONSIDERATIONS: RECOMBINANT DNA

Worldwide social upheavals in the 1960s reached the United States in the form of radical student movements, a hippie culture of drugs and liberated sex, and intensifying antiwar and antiestablishment confrontations. By the early 1970s, these powerful forces of change and protest had focused on opposition to the Vietnam War and on environmental concerns such as nuclear plants and carcinogens. Many of these movements were highly organized, had immediate access to the media, and gained considerable support from professionals and intellectuals, including scientists.

During these tense years of political crisis in the nation, the possibility of genetic manipulation or genetic engineering finally became a reality. In 1970, Paul Berg, a biochemist at Stanford University, along with his colleagues, began the lengthy and laborious project of grafting in a test tube an animal tumor virus SV 40 (simian virus) onto a laboratory version of a bacterium, *E. coli*, found in the digestive tract of mammals, including man. Such a hybrid might be useful in genetic probes and other experiments. It might also escape from the test tube and with its attached tumor virus enter a human being, conceivably causing disease. At that time, laboratory techniques were awkward, unregulated, and rapidly evolving. When one of Berg's graduate students went east to the Cold Spring Harbor Laboratory for a summer course, she mentioned the project to other biologists. Some were dumbfounded that, considering the potential biohazards of escaped tumor viruses, such work could be undertaken by a responsible researcher. Telephone calls and letters were exchanged with people in California to raise the question of risks. Berg decided to postpone the experiment, perhaps to call it off. Widespread agitation about the allegedly sinister role of scientific research in perfecting military weapons and in polluting the biosphere intensified the apprehensions of scientists as well as of the general public. Many biologists opposed war in general and the Vietnam War in particular. A good number of them leaned toward political radicalism and sympathized with some of the voices that questioned scientific research.

Here begins the ten-year story of recombinant DNA, one of the most clearly defined and heatedly debated episodes concerning the prudential aspects of scientific research in this century. In comparison, the huge disputes provoked by Galileo and Darwin appear to be primarily theological. I shall reduce the story so as to bring out those elements that pertain to our exploration of forbidden knowledge.*

By 1973, apprehensions about recombinant DNA had increased enough to justify two closed meetings of molecular biologists at the Asilomar conference center (near Monterey, California) and at New Hampton, New Hampshire, to discuss what steps to take. After the Gordon Conference in New Hampshire, the chairman represented a number of participants in a letter to the president of the National Academy of Sciences. This statement stands alone as a paragraph.

"Certain such hybrid molecules may prove hazardous to laboratory workers and to the public. Although no hazard has yet been established, prudence suggests that the potential hazard be seriously considered."

The letter proposed that a standing committee do just that, and most of the letter was published in *Science,* the most widely read

---

*Two excellent books give the full story. James D. Watson and John Tooze reproduced five hundred pages of original documents from scientific journals, private and official correspondence, and articles in the popular press. In *The DNA Story* (1981), the participants on both sides speak in full for themselves in their original words. It often makes exciting reading. In *Genetic Alchemy* (1982), Sheldon Krimsky (with David Ozonoff) provides a careful narrative history and analysis based on archival sources and many interviews.

scientific journal in the United States. Recombinant DNA experiments continued and were generally treated in the press as highly promising, close to miraculous. Almost exactly a year later, a truly blue-ribbon committee chaired by Paul Berg released its report to the NAS on "Potential Biohazards of Recombinant DNA Molecules." The eleven biologists, including James Watson and several other Nobel laureates, urged that until further evaluation and development of precautionary measures, "scientists throughout the world join with the members of this committee in voluntarily deferring the following types of experiments. . . ." Two such experiments were described. Published both in *Science* and in its British equivalent, *Nature,* in July 1974, the document (soon to be called the "Moratorium letter") was treated as headline news deserving editorial comment—often with references to Pandora's box and to tampering with nature's secrets.

The Berg letter also called for an international meeting on the subject. Organized primarily by Berg, the meeting followed with remarkable swiftness six months later. It was again held at Asilomar, and this time it was open to avid and uninformed reporters. It is revealing that the most comprehensive coverage of the conference appeared in the formerly radical organ of the sixties generation, *Rolling Stone* (June 19, 1975). "The Pandora's Box Congress," written by Michael Rogers, contained a gossipy but generally reliable and trenchant account of the four-day conference. In his conclusion, Rogers mentions the possibility, raised in the course of the wide-ranging debate, of "the creation of novel biotypes never seen before in nature"—that is, monsters, more politely referred to as chimeras or chimerical plasmids. Such artificial manipulation of the evolutionary process, Rogers suggests, "will represent as profound an expulsion from the Garden as the human intelligence has thus far managed."

Asilomar II voted to recommend the classification of experiments into four levels of risk, requiring increasing measures of physical containment. The report, approved somewhat hastily on the final day, included this quiet yet momentous proviso, which was sustained by a show of hands: "4. *Experiments to be deferred.* There are feasible experiments which present such serious dangers that their performance should not be undertaken at this time with the currently available vector-host systems and the presently available containment capability."

Neither of the two Nobel laureates present, Watson and Joshua Lederberg, favored the recommended regulations. Eighteen months later, in mid-1976, the National Institutes of Health issued guidelines for recombinant DNA research; legislation by Congress was ultimately unnecessary. Several times reviewed and extensively relaxed, these guidelines remain in force.

The discussions preceding and following Asilomar II represent the first instance in history of a major group of research scientists adopting voluntary restrictions on their own activities.* Commenting two years later on the astonishing self-restraint of the biologists, P. B. Medewar made a pointed and only partially appropriate comparison.

> *No literary folk have ever done as much. On the contrary: any suggestion that an author should not write exactly as he pleases no matter what offense he causes or what damage he does is greeted by cries of dismay and warnings that any such action would inflict irreparable damage on the human spirit and stifle forevermore the creative afflatus.*
>
> (*The New York Review of Books*, October 27, 1977)

Ethical and humanitarian considerations were invoked many times at the conference. I believe, however, that the decisive factors were

*In 1939, a group of refugee physicists in the United States began to discuss the need to keep the possibility of an atomic chain reaction a secret from Nazi scientists, who were already on the scent. Leo Szilard, Isidor Rabi, Enrico Fermi, Edward Teller, Eugen Wigner, and Niels Bohr met at various times in New York and Washington, D.C., to consider political and military justification for a voluntary suspension of the fundamental principle of openness in scientific research. Their efforts were short-lived. Before they made much progress with the Department of the Navy and the American Physical Society, Hitler invaded Poland in September 1939 and World War II began. Roosevelt's creation of the Manhattan Project imposed secrecy on all research concerned with atomic fission. This brief episode of voluntary restraints *on publication, not on research*, is well told in Richard Rhodes, *The Making of the Atomic Bomb* (1986).

prudential and legal. As *The DNA Story* demonstrates by its extensive facsimile version of the documents, a howling mix of voices contributed to this decision, made not democratically by the mass of citizens or their representatives, but by the molecular biologists themselves addressing governmental agencies responsive to them. As things turned out, the molecular biologists had overreacted and had to begin backtracking very soon in order to avoid excessive regulation. They had to put out a fire they had themselves started. Had they been prudent, or merely foolish and self-serving? From the many voices in this babble, I shall single out two participants because they contributed information and arguments significant enough to sway many minds and because those two voices reveal the deeper dynamics of the debate.

Although no one knew it in the early seventies, a young staff writer for *Science* carved out for himself one of the key roles in the recombinant DNA episode. Nicholas Wade had scientific training in England and worked first for *Nature* before moving to *Science* in Washington, D.C. Early in the story, in November 1973, Wade was assigned to follow up on the Gordon conference letter about hazards and to write a background piece on recorded mishaps and infections in biological research. He produced for *Science* a compactly documented six-column article entitled "Microbiology: Hazardous Profession Faces New Uncertainties." Wade's readable style and startling opening sentence must have earned him many readers. "Since the turn of the century, some 3500 cases of laboratory-acquired infections have been reported, more than 150 of which have resulted in death." In describing some of these cases, Wade quotes a number of scientists whom he apparently reached on the telephone. Granting the possible risks of working with animal tumor viruses, George J. Temaro of the National Cancer Institute went on to say: "My guess is that it's considerably less dangerous than smoking two packs of cigarettes a day." Wade did not leave it at that. Not everyone he spoke to was so complacent. "What if the guess is wrong? 'We're in a pre-Hiroshima situation,' says Robert Pollack of the Cold Spring Harbor Laboratory." Pollack was in effect declaring an emergency. The article also points out the inadequate training of many chemists and physicists switching into biology and using pathogenic materials for the first time.

Wade went on to write some twenty major stories on recombinant DNA in *Science* during the next seven years. Drawing on a background in history and philosophy of science, and developing wide acquaintance among molecular biologists, he could function a little like a television anchorman managing an explosive story, and also as a closet agent provocateur. His reporting affected the story he was reporting on. His constant feed of articles in *Science* and the book he published in 1977, *The Ultimate Experiment: Man-Made Evolution*, kept people judiciously informed about a debate often interrupted by wild exaggerations, loud demonstrations, and confrontational tactics. Wade was no Luddite, but his chapter entitled "The Dilemmas of Demiurgy" revealed strong reservations about "the ultimate technology."

The other voice I wish to single out is not that of a journalist writing for an influential review, but the collective voice of three lawyers brought in to address the second Asilomar conference in 1975 the evening before the final morning session of voting on recommendations.* By all accounts, these three speakers both woke up the participants after three full days of technical discussions and obliged them to turn in a troubling new direction. In their talks and the ensuing discussion, the lawyers presented three propositions of increasing consequence. First, decisions about laboratory experiments entailing potential hazards should be made with the participation of the laboratory workers, the surrounding community, the public, and such existing government agencies as the Occupational Safety and Health Administration and the National Institutes of Health. The scientists alone were not competent to assume all responsibility. Second, the sacred freedoms of inquiry, thought, and speech on which scientific research is founded do not include the freedom to do harm, physical or otherwise, to human beings. Third, if a plaintiff convinced a lay jury of the existence of harm, under existing law individual scientists could be held personally liable for large damages on the basis of professional negligence. The lawyers suggested that a set of responsible regulations for laboratories and kinds of experiments would help protect

*For the sake of economy I shall lump together the messages of these very different individuals: David Singer from a Washington, D.C., law firm and the Hastings Institution; Roger Dworkin from Indiana University; and Alex Capron from the University of Pennsylvania Law School.

biologists. They might also want to consider liability insurance, the lawyers added.

The shock and even fright that these remarks inspired in the conference participants surprised the lawyers. Roger Dworkin later remembered:

> *What a legal audience would have regarded as commonplace, elementary, and obvious, struck the distinguished scientists as novel, shocking, and frightening. Calling the researchers' attention to their potential liability induced a fear in them akin to a lay person's fear of virulent bugs crawling out of a laboratory.*
>
> (KRIMSKY, 141)

In his regular articles in *Science*, Wade alerted biologists to the potential risks of their work and their ethical responsibility to pause, at least, in their work before plunging ahead. The lawyers gave these prudential concerns a specific shape: damage suits.

The Asilomar II conference and the ensuing 1976 NIH regulations (including a moratorium on certain experiments) close act one of the story of recombinant DNA. Act two focuses on a dispute that started in 1976 when molecular biologists at Harvard University proposed to convert several rooms of the biology laboratories to level three of physical containment under NIH guidelines. Opposition among other biologists in the department rapidly became university-wide and involved medical school professors. Within a few months, the mayor of Cambridge, Massachusetts, had intervened in the dispute and carried it to the city council. That body imposed a three-month moratorium on all forms of recombinant DNA research at both Harvard and MIT. Meanwhile, testimony before the Cambridge Experimentation Review Board received wide national attention in the media as Nobel laureates took diametrically opposed positions. Incensed laymen produced hair-raising scenarios—such as that of an African dictator obtaining a cancer-producing "ethnic weapon" that he uses to terrorize his enemies, including the Western nations. A lengthy article on the debate by Arthur Lubow in *New Times*, "Playing God with DNA," quoted Tocqueville, Max Weber, Brecht, and—most effectively—the prominent DNA researcher Erwin Chargaff. Chargaff had become a goad to his colleagues. "The idea that science can make a better world is hubris." Chargaff's words elicited an eloquent response in the *New England Journal of Medicine.* In his regular column, "Notes of a Biology Watcher," Lewis Thomas criticized hubris as a code word used by anti-intellectuals, insisted that the debate over DNA should be confined to prudential considerations, and concluded that true hubris lies in pretending "that the human mind can rise above its ignorance by simply asserting that there are things it has no need to know." This form of hubris "carries danger for us all."

The events in Cambridge and impending congressional hearings on stringent legislation regulating recombinant DNA led to act three.

> *During the past year or so, virtually all of the scientists who were among the first to express concerns about certain kinds of recombinant DNA experimentation have come to believe that our earlier concerns were greatly overstated. At the time this issue was first raised, the techniques were new.... However, while the anxieties of the public and of some members of Congress have been increasingly aroused, the work has proceeded without adverse consequences in many dozens of laboratories around the world.*

In October 1977, Stanley Cohen, a medical doctor and research geneticist at Stanford University, could make that statement to a congressman on the basis of "no harm to humans or to the environment" in four years of recombinant DNA experiments. Gradually and almost reluctantly, biologists pulled themselves together to turn back the forces of fear that they had themselves unleashed. Not the fear of risks and negligence suits, but the fear of laboratories being shut down drove them now.

These disputes, practical and principled, over DNA experiments seared their way deep into the conscience of some concerned scientists. A respected and articulate molecular biologist at California Institute of Technology, Robert Sinsheimer, published in 1969 an article in *Engineering and Science* favoring "a new eugenics." "For the first time in all time, a living creature understands its origin and can undertake to design its future....

"The new eugenics would permit in principle the conversion of all the unfit to the highest genetic level" (quoted in Kevles and Hood, 18, 289).

This freewheeling attitude toward genetic engineering had changed completely by the time Sinsheimer became chancellor of the University of California at Santa Cruz in 1977. Two years earlier, in an article entitled "Troubled Dawn for Genetic Engineering," he called for limits on rDNA research with a rhetoric that goes far beyond prudential concerns. "It is no longer enough to wave the flag of Galileo.

"Rights are not found in nature. Rights are conferred within a human society and for each there is expected a corresponding responsibility" (Watson and Tooze, 55).

In an interview published in *Science* in 1976, Nicholas Wade quotes Sinsheimer as saying: "To transgress [the natural barrier between eukaryotes and prokaryotes] in hundreds of laboratories throughout the world is to risk unpredictable—and irreparable—damage to the evolutionary process." (Watson and Tooze, 147). *Daedalus*, the journal of the American Academy of Arts and Sciences, asked Sinsheimer to contribute to its 1977 issue on "The Limits of Science." He entitled his essay "The Presumptions of Science" and opened with this abrupt question: "Can there be 'forbidden'—or, as I prefer, 'inopportune' knowledge?" Obviously troubled by essentially metaphysical and moral questions about the consequences of his own profession, Sinsheimer later underwent a second conversion. In May 1985, he organized in Santa Cruz the first high-level workshop to consider the technical feasibility of what would soon be called the Human Genome Project. Such a scheme would make no sense unless the data it amassed could be employed in extensive rDNA research and therapy. Sinsheimer's tergiversations testify to a period of self-doubt in the 1970s among molecular biologists, a reaction based principally on fluctuating prudential concerns. When the risks failed to materialize, the doubt gave way to renewed confidence and enterprise.

By 1980, the tide of opposition to rDNA experiments turned. The Supreme Court approved a disputed patent submitted by Stanley Cohen and others on "a process for . . . biologically functioning molecular chimeras." Five judges out of nine recognized the patentability of man-made forms of life. In an address to UNESCO the same year, Pope John Paul II spoke with concern about "genetic manipulations and biological experiments" as "destructive of the true dignity of human life." That was a different issue, however, from biohazards.

What does this seven-year episode reveal about the risks and benefits of scientific research? One interpretation insists that molecular biologists themselves, by crying wolf out of ignorance and spinelessness, slowed down rDNA research quite unnecessarily just as it began to gather momentum. In many respects, James Watson and others with his confidence were right from the start. But they based their estimate of risks not on documented knowledge but on informal hunches about the behavior of bacteria domesticated in the laboratory. A dispassionate observer might well conclude that the pause in certain categories of research with new hybrids, especially "shotgun" experiments, represented responsible behavior. It could be seen as a reasonable response to Mary Shelley's clarion early warning 150 years before in *Frankenstein*. The lesson of the rDNA story is *not* that we should never worry in the future about prudential concerns, even when we reach an uncertain boundary surrounding forms of human life. The lesson is that this episode provoked a public airing of questions not previously raised outside of scientific circles. The debate culminated in a flexible set of guidelines and the formation of an oversight panel, the Recombinant DNA Advisory Committee (RAC). It continues to function actively, serving both science and the public interest. We all deplore the proliferation of bureaucracy and regulations. But in this case, the RAC has gone a long way toward preventing an outbreak of the Frankenstein syndrome among the more fanatic genetic engineers. The rDNA story recapitulates the dilemmas and paradoxes of scientific research and underscores prudential considerations without obscuring others.

Legal Considerations: Buck v. Bell *and Justice Holmes*

The third category of circumstances that could justify constraints on scientific inquiry reaches beyond practical and prudential concerns into the maze of the law. Where, for example, does the march of science interfere with individual rights? Does the present court system tend to give "scientific evidence" more weight than it deserves?

My example is drawn from the decade of the 1920s in the United States, when the great advances in science still provoked stubborn and organized resistance. In 1925, the state of Tennessee set itself up like the Victorian matron I quote on the second page of the Foreword and passed a law to prohibit teaching any but the biblical account of man's creation. The ensuing Scopes trial pitted William Jennings Bryan of Nebraska, the great populist reformer, orator, and religious fundamentalist, against the Chicago lawyer for underdog causes, Clarence Darrow, who was defending the biology teacher John Scopes. The nation's first coast-to-coast radio hookup broadcast Darrow's unsparing cross-examination of the three-time presidential candidate. In what Mencken described as the atmosphere of a blast furnace, Darrow asked Bryan if he had studied other religions and the history of ancient civilizations to verify his literal interpretation of the Bible. Bryan's response makes a pitifully benighted argument for avoiding or forbidding certain kinds of knowledge. "I have all the information I want to live by and die by." Legally, Bryan was on the winning side, but Darrow made considerable inroads for evolution and against creationism.

My principal example is not the Scopes trial, but a related case from the twenties. It reached the Supreme Court and invoked the new science of eugenics. Widely endorsed by biologists and prominent philanthropists, eugenics sought to develop an improved breed of the human species by restricting births from the "unfit" and by encouraging births from the most fit. And the case also concerns one of our most renowned justices.

Toward the end of his long career, Supreme Court justice Oliver Wendell Holmes ignited a great blaze in the drafty mansion of American law with his liberal free-speech opinions of the early 1930s. Just a few years earlier, he wrote a majority opinion for the Court from which there was only one quiet dissent. *Buck* v. *Bell* (1927) causes evident embarrassment to Holmes' admirers and biographers. We know from his letters how proud Holmes himself was of the decision and of the resolute way in which he wrote it up.

After World War I, several states passed sterilization laws for the unfit or the feebleminded based on eugenicist principles. According to Daniel Kevles, close to nine thousand people in all were sterilized by 1928, twenty thousand by the mid-1930s. Opponents mounted a strong attack on the practice as unscientific and ineffectual. State courts began declaring such laws unconstitutional on various grounds. After three years of appeals, the Supreme Court heard *Buck* v. *Bell*, a case in which all previous decisions sustained the Virginia sterilization statute and its application in these circumstances. Carrie Buck, the illegitimate child of a feebleminded mother and adopted at age four, became pregnant at age seventeen and was committed to the State Colony for Epileptic and Feebleminded in Lynchburg, Virginia. The superintendent petitioned to have Carrie sterilized. Carrie's attorney fought the petition and carried the fight to the courts. The Colony, represented by Aubrey Strode, a lawyer of eugenicist convictions and author of the Virginia statute, prepared a case based in part on expert testimony from Harry Laughlin of the Eugenics Records Office in Cold Spring Harbor. They declared that Carrie and her mother were feebleminded and promiscuous and that Carrie's baby was also feebleminded. Another expert declared, "The blood is bad." Carrie's attorney pleaded due process, cruel and unusual punishment, and equal protection.

In Washington, D.C., in 1927, the case provoked little interest and was quickly settled in favor of sterilization, a procedure that would allow Carrie to be discharged from the Colony. Justice Holmes' opinion accepted both the facts of the case as presented by the Colony and the claims of the eugenics experts about the transmission by inheritance of insanity and imbecility. Due process had been observed, he found, and "her welfare and that of society will be promoted by her sterilization." The following sentences have become celebrated because they recognize no special protection of procreation in the Constitution.*

> *We have seen more than once that the public welfare may call upon the best citizens for their lives. It would be strange if it could not call upon those who already sap the strength of the State for these*

---

*Thurgood Marshall in *San Antonio* v. *Rodriguez* (1972) cites *Buck* v. *Bell* as the "initial decision" to that effect. *Roe* v. *Wade* (1973) made substantial use of that lack of protection for procreation in modifying the law on abortion.

> *lesser sacrifices, often not felt to be such by those concerned, in order to prevent our being swamped with incompetence. It is better for all the world, if instead of waiting to execute degenerate offspring for crime, or to let them starve for their imbecility, society can prevent those who are manifestly unfit from continuing their kind. The principle that sustains compulsory vaccination is broad enough to cover cutting the Fallopian tubes.... Three generations of imbeciles is enough.*
>
> (*Buck* v. *Bell*)

Negative eugenics enforced by compulsory sterilization gives us pause even when scrappily defended by Holmes. This is Faustian science. The words that should trouble us in the passage are "manifestly unfit." Holmes and the Court failed to probe into the facts and theories adduced by the experts. Robert Cynkar and Stephen Jay Gould have published new evidence that Carrie was probably raped by a member of her adoptive family and then put away to hush things up. When released after sterilization, she lived a normal and dignified life. No sound test or examination ever established that mother or daughter was imbecile or feebleminded. Furthermore, the contention that the feebleminded inevitably breed feebleminded offspring—the premise on which the whole case rested—arose from unreliable scientific information and should have been questioned in Carrie's defense. But her lawyer was a friend of Strode's, and he was brought in to stage this test case, which was intended from the start to go to the Supreme Court and to vindicate eugenicist sterilization. Holmes missed the opportunity to affirm the human right to bodily integrity and to reproduction except in the event of "clear and present danger" to the community—Holmes' great principle protecting individual rights. Instead, he took the reformer's course of promoting the public welfare in the name of flimsy eugenicist theories not subjected to scrutiny.

There is no simple way of preventing half-baked science from promoting its claims and from influencing court cases. We have become more skeptical of expert testimony than the Court was in 1927. But if anything, legal concepts growing up alongside of science entangle us more than ever. One of the most forthright outbursts against sterilization was made by governor Bibb Graves of Alabama in 1935 in vetoing a sterilization law. The statute, he said according to *The New York Times* (September 5), would punish many women "who have committed no offense against God or man save that, in the opinion of the experts, they should never have been born."*

Half a century later, prenatal screening, not sterilization, opens up new frontiers of litigation. It would be difficult to argue that the legal climate has improved. Now we can sue doctors or health-care providers or genetic laboratories if they fail to provide full and reliable information about a pregnancy that leads to the birth of a seriously impaired child. State courts have already decided cases of *wrongful birth*, in which the parents bring the action, and of *wrongful life*, brought by the child (by proxy) claiming it should never have been born and seeking compensation for pain and suffering undergone as a result—even for being deprived of a full life span. In such circumstances, words like *compensation* and *damages* become a ridiculous mockery. Once again we are throwing ourselves into the hands of "experts" and asking that their knowledge remove all risk of inherited disease. In a 1980 wrongful-life case, one California judge stated: "The certainty of genetic impairment is no longer a mystery" (Kevles, *In the Name of Eugenics*, 293). He went much too far. Prenatal tests can distinguish only a few single-gene inherited and incurable conditions for which we do not have the resources to screen the entire population of this country, let alone the world. Litigation on wrongful life and birth probably does not improve medical standards enough to justify the enormous costs it imposes on health care and on the public psyche.

Now the perplexities of sterilization and screening cases in law do not justify an attempt to shut down research on prenatal and carrier testing and on genetic research in general. But, as I pointed out earlier, a geneticist as prominent as Robert Sinsheimer could warn us against the "risk of unpredictable—and irreparable—damage to the evolutionary process." Montaigne and Pascal used the word *presumption* to describe this pursuit of curiosity beyond our

---

*Graves' sentence inverts the revolutionary reasoning of Figaro. In Beaumarchais' play, the servant protests his noble master's privileges, earned by no greater achievement than "to have been born."

lagging capacity to deal with the results. Psyche found out the truth about her nocturnal lover, but the consequences destroyed what happiness she had and the balance of her existence. I should not have to belabor the connections between the first part of this book and the second.

As we approach the end of the twentieth century, sterilization no longer represents a highly disputed issue in public policy, even though earlier debates and cases remain important precedents. Today, both in Europe and in the United States and Canada, guidelines or regulations are in place that carefully oversee any gene therapy of somatic cells (that is, of genes in a patient's ordinary cells in order to yield modifications that will not be inherited by the patient's offspring) and that categorically prohibit gene therapy of germ cells (that is, of a patient's egg or sperm in order to yield modifications that will be inherited and will thus enter the human germ line). The distinction is fundamental to the present discussion. On July 22, 1982, *The New York Times* published an editorial entitled "Whether to Make Perfect Humans." It called for careful discussion not only of who should decide such matters but also what is at stake. "There is no discernible line to be drawn between making inheritable repairs and improving the species."

The next year, the activist Jeremy Rifkin, author of *Algeny* (1983), an impassioned challenge to evolutionary theory, organized an astonishing coalition to oppose intervention in the germ pool. Twenty-one Roman Catholic bishops, at least two fundamentalist evangelists (Jerry Falwell and Pat Robertson), Nobel laureate George Wald of Harvard, and several Protestant and Jewish religious leaders held a press conference to issue a resolution opposing any attempt "to engineer specific traits into the germ line of the human species."

Both the *Times* editorial and Rifkin's caucus opposed any interference with the invisible hand—of natural selection, or of a divine being—that has brought the human race to its present juncture. Fortunately, in the United States and in Europe we already have in place institutions and procedures through which such crucial decisions will be made by scientists and laymen working together. Furthermore, the scientific community has a strong international cohesiveness that may help to prevent such decisions from fragmenting into parochial or national solutions. Should we, then, try to apply reasonable principles and regulations to all genetic research everywhere in the name of a commonsense respect for human nature and of our reluctance to tamper with it?

For the immediate future, I would answer yes. Furthermore, at this point in the discussion, prudential and legal considerations have blended into moral choices. And the long view complicates matters even more. At the end of *The Human Blueprint: The Race to Unlock the Secrets of Our Genetic Script* (1991), the chemist Robert Shapiro distinguishes issues of species survival, such as preventing nuclear war and preserving the ozone layer, from personal choices, such as "editing our genetic text," which includes our germ cells. Shapiro favors unregulated diversity of genetic research in humans and mutual toleration of those programs more than he favors cooperation among communities.

> *For moral reasons, or because they value themselves as they are, some cultures may wish to keep their germ lines inviolate. Others may decide to make modifications, but only to eliminate genetic diseases. Yet others may allow individuals to introduce "improvements" as they choose. The options permitted will vary from place to place.*
>
> (372)

There are several things seriously wrong with Shapiro's position. Most important, we do not come close to knowing enough about the complex correlations among genes, human development, disease, behavior, and evolution to make sound scientific or moral decisions about intervening in that process. Because of pseudogenes, silent genes, remote-control elements, and the like, we cannot define precisely the nature of "a gene." We cannot even be sure of identifying an "improvement." What we do know tells us that almost no identifiable cause in genetics has a single effect, and vice versa—particularly over time. Gene functions change. And nothing justifies so simplistic an analogy as "editing our genetic text." DNA does not resemble printed prose that can be edited in one spot with no side or subsequent effects. That kind of misleading comparison makes us impatient to try experiments when we should be learning patience and looking into the long-term effects of genetic intervention in lower organisms.

A fine antidote to Shapiro can be found in *Physician to the Gene Pool: Genetic Lessons and Other Stories*, a probing memoir by James V. Neel. First a molecular biologist, then a practicing physician, Neel finally became a population geneticist. He directed two studies on the radiation effects of the atomic bombs dropped on Japan and spent several years studying the genetic makeup of an isolated Indian tribe in the Amazon valley. Out of his varied experience, Neel draws the sober conclusion that what we have learned medically and genetically in the past fifty years has brought us to a serious crisis. For what we *do* know allows us to prolong individual life in ways that aggravate the largest problem facing us: population outgrowing resources. And what we do *not* know tempts us to try further steps like somatic gene therapy before we have ascertained its collateral and long-range consequences. He advises more and better education about reproduction, diet, exercise, and the like and less government-sponsored research devoted to gene therapy. In one particular area, Neel takes a strong stand. He cites his own remarks on leaving the Council for the National Institute for Aging. In promoting research to prolong life, particularly by intervening in the genetic process of "aging switches," the council is serving objectives antithetical to social health. For we cannot now adequately sustain the number of aging citizens and should not encourage more. Neel gave the National Institute for Aging the advice of someone concerned not with the special interests of the aging but with the general welfare.

> *Given conformity with ethical standards, there can be no forbidden research in a democratic society, but I would assign a low priority to research directed at altering the functioning of the "aging switches," as contrasted to research on diseases of old age (which if successful of course implies some prolongation of average life span).*
>
> (386)

One response to Neel here is that we should maintain and sometimes enforce the category of forbidden research. Still, Neel is both scientifically sound and eloquent in opposing attempts to carry out supposedly therapeutic procedures on the basis of inadequate knowledge of genetic mechanisms. Neel's warnings closely parallel those that might have been given to Justice Holmes in the *Buck* v. *Bell* case, warnings against unreliable testimony coming from the Eugenics Records Office. For Holmes's confident majority opinion was based on faulty "expert" testimony.

Moral Considerations: Himmler's Lebensborn

The step from legal to moral considerations leads us to questions of even greater complexity. I shall illustrate them with a case history arising from science under totalitarianism. The best way to approach that case lies through a little more background in genetics and eugenics.

Gregor Mendel, the Austrian monk-scientist, was the Adam of genetic science. His twenty-year experiment with garden peas yielded the basic laws of inheritance. Nothing could show greater reverence than those descriptions of the behavior of dominant and recessive traits in a domesticated legume. But reverence does not drive all scientists. In the 1920s, a brilliant geneticist at the University of Texas, Hermann Muller, discovered how to override those laws by using X rays to induce mutations in fruit flies. He later received the Nobel Prize for this work.

A cranky, somewhat unstable man, Muller was said by one of his friends to have traded in the three Rs for the three Ss: science, sex, and socialism. He had picked up these loyalties as a graduate student in New York and as a research scientist in Leningrad. He courageously condemned Lysenkoism in the Soviet Union, yet he saw the future of society in terms of Leninist socialism, while deploring the competitiveness of American capitalism. More than any other American scientist, Muller represents a multiple shift in biology in the 1920s and 1930s: focus no longer on the organism or cell but on the *gene* as the unit of life; understanding the gene as carrying information, a crucial code to be cracked; and welcoming a Frankenstein-like manipulation of the processes of life in order to achieve particular social goals. His 1935 book, *Out of the Night*, preached "entelegenesis," or eugenic breeding by the use of artificial insemination and the creation of test-tube babies. The book made little impression in the United States and then received resounding endorsements in England from G. B. Shaw, C. P. Snow, J. B. S. Haldane, and Julian Huxley.

They found compelling Muller's most progressive ideas. "How many women, in an enlightened community devoid of superstitious taboos and of sex slavery, would be eager and proud to bear and rear a child of Lenin or of Darwin! Is it not obvious that restraint, rather than compulsion would be called for?" (122). Shaw and Huxley did not hesitate to promote these ideas of separating human reproduction from both sexual love and family bonds. Eugenics could reach a long way into social reform.

In 1939, Muller decided to promote his ideas by distilling them into a fifteen-hundred-word statement with six points. He persuaded twenty-two distinguished British and American biologists to sign it.* The document was issued at the Seventh International Congress of Geneticists in 1939 in Edinburgh, appeared in *Nature* (September 16), and has come to be known as "the geneticists' manifesto." Its utopian goal of improving the world's population rests on a promise that calls for "control" of broad areas of human life: ". . . both environment and heredity constitute dominating and inescapable complementary factors in human well-being, but factors both of which are under the potential control of man and admit of unlimited but interdependent progress."

To attain this goal in heredity, "some kind of conscious guidance of *selection* is called for. . . . This in turn implies its socialized organization." Lofty idealism yields gradually to dismaying naïveté as biology and politics join hands in the following passage. An alert reader will hear the debate between socialism and capitalism.

> *The most important genetic objectives, from a social point of view, are the improvement of those genetic characteristics which make (a) for health, (b) for the complex called intelligence, and (c) for those temperamental qualities which favour fellow-feeling and social behaviour rather than those (today esteemed by many) which make for personal "success," as success is usually understood at present.*
>
> *A more widespread understanding of biological principles will bring with it the realization that much more than the prevention of genetic deterioration is to be sought for, and that the raising of the level of the average population nearly to that of the highest now*

---

*The list includes scientists of lasting fame: C. D. Darlington, J. B. S. Haldane, J. S. Huxley, J. Needham, Theodosius Dobzhansky, C. H. Waddington.

> *existing in isolated individuals, in regard to physical well-being, intelligence and temperamental qualities, is an achievement that would—so far as purely genetic considerations are concerned—be physically possible within a comparatively small number of generations. Thus everyone might look upon "genius," combined of course with stability, as his birthright. As the course of evolution shows, this would represent no final stage at all, but only an earnest of still further progress in the future.*

Progress will ensue if the state, representing the people, can take over natural selection and give us our birthright of genius. The manifesto calls for genetic research "on a much vaster scale as well as more exact." One wonders if Muller and his cosigners had read *Brave New World*, published seven years earlier by Aldous Huxley, and if they suspected what was going on under their noses in Nazi Germany. The last sentence of the manifesto refers vaguely to the need to overcome "more immediate evils" before reaching "the ultimate genetic improvement of man." The sentence probably alludes to the abuse of science in all societies. But did the nonaggression pact between Hitler and Stalin in August 1939, concluded precisely as the eugenicists' manifesto was being drafted, preclude condemnation of what was happening in Germany? I doubt if even Muller would have followed the Communist party line to that extent. Ignorance is a more plausible explanation.

In any case, our eyes are now open to what was going on under Hitler in the name of genetic improvement. As support for mainline eugenics began to wane in the United States and Great Britain in the 1930s, it became state policy in Germany. I am referring not only to the horrors of negative eugenics, for which we have had to invent the word *genocide* and about which I shall not write, but even more to little-known social experiments in positive eugenics to breed the fittest and best-endowed Germans.

The most telling account of the latter has come from a survivor with the persistence to penetrate the conspiracy of silence.* After

---

*See Marc Hillel and Clarissa Henry, *Of Pure Blood*. The investigations of these two French journalists have been extended and enlarged in the scholarly work of Georg Lilienthal.

the losses of World War I, Germany felt a profound urgency to increase its population and its birthrate. At the same time, persuaded by a series of racist thinkers whose message became state doctrine under the Nazis, the ideal took the shape of purifying the German people by "racial hygiene"—state-approved marriages and experiments with artificial insemination. Put into practice under Hitler, these policies led to some strange contradictions. The effort to lure women back into the home and productive family life, summed up in the slogan *Kinder, Küche, Kirche* ("Children, Cooking, Church"), yielded in the late 1930s to a program of weakening the family unit and encouraging illegitimate births of children, who would then become charges of the Third Reich. Accordingly, Heinrich Himmler lived openly with a mistress and did not hide the children he had by that union. Martin Bormann did the same. Fidelity to the family unit might interfere with prolific breeding.

Head of the entire state police, the Gestapo and SS, probably the inventor of the gas chamber, a major organizer of the Final Solution, and Hitler's major rival, Heinrich Himmler also founded and consistently supported a sinister institution with a beautiful name: *Lebensborn*, meaning "fountain of life." After 1935, these maternity homes gradually assumed functions other than simply helping unwed mothers. They became screening centers from which the physically unfit and racially unwanted were sent away for "disinfection" or "resettlement." Both terms meant extinction. Often such homes were established next to the barracks of Himmler's racially elite SS corps. The encouragement given to this kind of frequentation and breeding gave these homes the reputation of being houses of prostitution or stud farms for the SS. A certain number of ghoulish racial and eugenic experiments were carried out under the auspices of these institutions. This arrogant state attitude toward the breeding of Nazi subjects led to an astonishing incident in January 1943.

Because of a series of stirring anti-Nazi leaflets distributed in Munich and other signs of disloyalty, students at the university (many in uniform) were summoned to a mass meeting by Paul Giesler, *Gauleiter* of upper Bavaria and group commander of the storm troopers. Giesler fulminated against the male students for slackness and malingering in the war for the Fatherland. Then he turned to the female students and adjured them "to make an annual contribution to the Fatherland of a child—preferably a son." He continued with a leer: "If some of you girls lack sufficient charm to find a mate, I will be glad to assign you one of my adjutants for whose ancestry I can vouch. I can promise you a thoroughly enjoyable experience" (quoted in Richard Hanser, *A Noble Treason*, 220). Giesler's performance inspired foot-shuffling, murmurs, heckling, whistles, women and men walking out, and, finally, a full-fledged protest. SS men at the doors could not control the demonstration, which spilled out into Ludwigstrasse. Arm in arm, the students sang and chanted slogans in the only open display of political defiance that ever occurred in Nazi Germany. A month later, three uncowed student distributors of the pamphlets, the leaders of the White Rose Resistance group, were beheaded by guillotine on Himmler's explicit orders.

When the Nazis invaded and occupied other countries in Europe, particularly those considered Aryan, they began to claim for the Third Reich children fathered by German troops. A few *Lebensborn* homes were established outside Germany. In some cases, mothers were kidnapped with their children. As time went on, a vast operation of baby snatching to supply the foreseeable need for manpower was carried out under Himmler's direction. The children were selected according to strict physiognomic rules and measurements based on race. The undesirables were sent to labor camps or simply eliminated—by allowing them to freeze to death, for example. Severe discipline, brandings, and injections to hasten maturity were common. The preferred children were separated from their mothers, indoctrinated with Nazi propaganda, adopted by German families, and repossessed at puberty by the state.

By this large-scale breeding and resettlement project in positive eugenics, Himmler and his associates exploited scientific research for social purposes—to augment the German birthrate and to monitor the racial purity of the population. According to Hillel and Henry, these goals were never achieved, despite Himmler's personal interest. Little wonder. Not only did this mad Faustian raid on scientific knowledge violate the sanctity of human life; it relied on erroneous science. In the last years, the *Lebensborn* homes became pockets of administrative graft with decreasing central control. As one might expect, the children, without parents and subjected to virtual incarceration, often turned into animallike creatures unequipped to contribute even to the Nazi cause. Himmler committed suicide in 1945, and at Nuremberg, investigation of the Race Settlement Office and the *Lebensborn* program did not probe adequately into the record of those organizations. As many as 200,000 kidnapped children, particularly from Poland, were never identified and never repatriated.

Under different circumstances, without the goal of racial hygiene, could any extensive project of positive eugenics for improving the human gene pool accomplish its ends in accord with human dignity and freedom? Let us confine the immense question to the presumably enlightened proposals of Muller and his distinguished colleagues in 1939. They called for "socialized organization" to achieve three "genetic objectives": health, the complex called intelligence, and "fellow-feeling and social behavior." One can imagine many schemes, some without breaking up the family unit, by which these goals might be approached. But we have learned to be suspicious of utopias. The principal obstacle takes a form evident to most people. The implementation of any such scheme requires that questions about the nature of "intelligence" and "fellow-feeling and social behavior" be satisfactorily decided in advance. For philosophical, moral, and human reasons, in a free society those questions will long remain in dispute. If they are decided by a dominant minority and incorporated into a social system, other citizens may lose the freedom and dignity essential to the human condition.

Still, should we not now take a few small steps toward steering our own course on the basis of our limited knowledge? At least we might reduce a few hereditary diseases. But some optimists want more. Muller and the twenty-two biologists who envisioned "raising the level of the average population nearly to that of the highest now existing in isolated individuals . . . within a few generations" were favoring a scientific endeavor most of us would resist. They wanted to move too fast in a direction that would shrink the gene pool and dehumanize society.

In the fifteenth century, Nicholas of Cusa constructed a theology based on *docta ignorantia*—"wise ignorance." I have examined the word T. H. Huxley devised for the modern scientific version of wise ignorance: *agnosticism*. What we know informs us increasingly about what we do not know. Given such limitations, can we formulate ultimate questions? Yes, tentatively. Those questions must address the origins, composition, basic processes, and direction of the universe, inanimate and animate. One further question subsumes the others: Are these the right questions? The fact that we remain ignorant of how to pose the final questions (even if we have a history of powerful answers to them) should encourage in us an attitude of reverence and wonder toward the world. If we all insist on hearing the Sirens' song, the chances are that many of us will not take the precautions Odysseus did to protect himself and his crew.

That fundamental ignorance in the midst of our flourishing knowledge provides the moral basis for moving slowly in the darkness and for resisting both the pull of reductionism and the lure of grand theory. We have little reason to be proud of our waverings, particularly in recent years. The answers given to ultimate questions by influential representatives of educated society have changed by 180 degrees within my memory.

> *Give me a dozen healthy infants, well-formed, and my own specified world to bring them up in and I'll guarantee to take any one at random and train him to become any type of specialist I might select—doctor, lawyer, artist, merchant-chief, and, yes, even beggarman and thief, regardless of his talents, penchants, tendencies, abilities, vocations, and race of his ancestors.*
>
> (John Watson, *Behaviorism*, 82)

Watson's *Behaviorism* (1924) dismissed both subjective states and hereditary factors and established social environment as all-powerful in human development. Ten years later, Ruth Benedict reinforced the influence of behaviorism over social policy with her *Patterns of Culture*. "For better or for worse, man's solution lies at the opposite pole [from biological mechanisms]. Not one item of his tribal social organization, of his language, of his local religion, is carried in his germ line" (12). Now, two generations later, we have another Watson—James—to proclaim the opposite. Having found the structure of DNA with Francis Crick, Watson looks to molecular biology for the true wisdom: "The genetic message encoded within our DNA will provide the ultimate answers to the chemical underpinnings of human existence" (*Science*, 6 April 1990).

A generation of geneticists has followed James Watson and claimed our attention. And for twenty years now, the new synthesis of sociobiology has been affirming and elaborating Watson's vision. E. O. Wilson defines sociobiology as "the systematic study of the biological basis of all social behavior." We should keep careful track of intellectual fashions in science, above all in our own time. Must I believe James Watson over John because James came later? Or because genetics has stood up better than behaviorism? And has it?

The moral of the story of Himmler's *Lebensborn* program and of the two Watsons points us, I believe, toward a wise agnosticism, toward the Angel Raphael's injunction to Adam and Eve: "Be lowly wise."

Mixed Considerations: The Human Genome Project

As the end of the twentieth century approaches, some disputes over the relations between genetics and society seem to have been laid to rest. Institutions no longer seek sterilization orders for the feebleminded, even though the legislation may remain on the books. We no longer worry constantly about epidemic diseases being caused by experiments with recombinant DNA. Only a few communities like Singapore have been trying out social policies based on eugenics, and none so racist and inhuman as the *Lebensborn*. But other disputes have arisen.

Surrounded by controversy since first suggested in the mid-1980s, the Human Genome Project proposes not only to map the 100,000-odd genes contained in our twenty-three pairs of chromosomes but also to sequence the 3 billion nucleotide bases that form the DNA of the complete human genome. For half a century since the Watson-Crick model, scientists have been aware of the crucial nature of this information; for a decade, technology has existed to make sequencing feasible. It is true that at least 90 percent of the bases in the human genome seem to code for nothing. Those bases form puzzling introns between the significant exons. They can hardly be, as the more poetic geneticists once thought, "junk" or "garbage." It is also true that the laboratory techniques of the HGP pose no evident prudential risks, that it will produce data useful in genetic research, and that, in spite of the inflated vocabulary used by its boosters *(adventure, wager, conquest, breakthrough)*, the HGP represents a large-scale, expensive, yet unchallenging piece of ordinary science. Should we get excited about it?

Without doubt, the HGP will contribute to techniques already used to predict the onset of certain inherited diseases. Mapping our genes and sequencing our DNA will improve and refine the way we screen an unfertilized egg, a fertilized egg before implantation ("pre-embryo"), prenatal fetuses, the newborn, and adult carriers in order to identify and respond to inherited disorders. At present, newborns may be tested for over ten disorders; we have located about eight hundred on the chromosomes; we have identified five thousand disorders as being inherited. The rush of money to support genetic workers and laboratories should have beneficial side effects. Even without the HGP, the complete sequence would slowly be filled in.

In spite of these advantages, a fairly stubborn and vocal opposition to the project has grown up. For critics both within and outside the scientific community, the HGP has come to represent a good cause gone wrong, a technology that has overwhelmed its inventors and become an end in itself, science deluding us into false hopes.

The objections fall into five overlapping categories. The next few pages are the most technical in the book. I believe my argument requires close scrutiny of this case history, through which we are living.

1. The HGP is an immense boondoggle, more ephemeral and less practical than an emperor's new palace and correspondingly lucrative to certain categories of workers. The government has supported it on the recommendation of those who will benefit most from it: researchers in genetics and molecular biology, administrators and bureaucrats in related fields, some doctors and medical personnel working in gene screening and therapy, and commercial firms and entrepreneurs. Many active promoters could be accused of conflict of interest and would, in legislative or judicial proceed-

ings, have to disqualify themselves. A hard-nosed objective evaluation of the HGP by a committee of outside scientists and medical personnel and laymen would recommend many revisions in the procedure, priorities, funding, and oversight.

2. The HGP represents an instance of large-scale centrally administered science, federally funded without adequate justification. The Manhattan Project and (more technologically than militarily) Project Apollo responded to challenges from a foreign power threatening our security. Both were crash programs mobilizing our best scientific minds and immense resources. No comparable crisis motivated those in federal government to initiate accelerated research in a specific area of biology. Such concentration of support upsets the balance of federally funded scientific investigation.

3. The HGP is bad science on two principal grounds. Even though well over 90 percent of the genetic code is the same in all human beings, there is still no standard genome, no single sequence of the four nucleotide bases that defines "human." Our species is defined both by the billions of base pairs we share with one another, and, equally, by the differences (polymorphisms) between individual genomes that provide the variety of human bodies and minds. The HGP will sequence the bases not of any one individual but of a random composite, or mosaic, of individuals according to where the work is farmed out and whose DNA a laboratory happens to pick out of its library. The essential operation of comparing sequences and genes of different individuals and correlating them with the lives of those individuals will not be undertaken by the HGP. It will pursue the anonymous and routine work of recording every nucleotide in a nonexistent straw man or collage of humanity. It will take as its central task what would be the eventual by-product of more modest research designed to focus on certain promising loci or genes in a number of individuals and the variation of those genes' expression in actual lives. Central authority has produced a reversal of priorities.

Second, the HGP tends to promote a single-factor explanation at a low order of magnitude for biological phenomena. It neglects larger and more complex units like the cell, organ, organism, and species. Horace Freeland Judson, a fine historian of science, represents this tendency in his affirmation of the word *ultimate.* "The sequence [of base pairs in human DNA] has often been called 'the ultimate map,' but this is a considerable semantic mistake—for the sequence is, rather, the ultimate territory of the genome project. And maps are never the territory mapped" (quoted in Kevles and Hood, eds., *The Code of Codes*, 78).

Judson's semantic and scientific bias here looms larger than the grain of truth the passage refers to. In an existing individual, the base sequence forms a key part of the total organism and, yes, part of the territory. But Judson has put on blinkers. The DNA molecule, inert, incapable of reproducing itself or of producing anything else, can be called "ultimate" only by a limited point of view that favors tiny size and whatever registers as code or information. Many elements of the organism—for example, neurons, hemoglobin—could be called "ultimate" because equally essential. None can be dispensed with, not even muscle fibers. By concentrating our attention and our resources on one aspect of the organism, the smallest and most blindly mechanical, the HGP distracts us from a proper or full understanding of ourselves. We have a severe word for this approach: *reductionism*.

4. The HGP spreads the illusion of offering a panacea. Almost every official document employs an expansive, sometimes breathless style, holding out the promise of a new world created by the application of genetic knowledge. Commercial and journalistic writings tend to the hortatory. Insofar as these publications rely on a logical argument, usually unstated, it runs along these lines.

Like all living organisms, human beings are governed by their genetic material, which encodes our life cycle as completely as what used to be called "fate" or "destiny." Having now discovered the fine structure and pervasive functions of DNA, we can begin to diagnose the source of many of our inherited ills and to design prenatal procedures (selective abortion, implantation, and in vitro fertilization) to eliminate or avoid them. Gene therapy that changes the DNA in a living person has been attempted with uncertain results to treat an immune-deficiency disorder. Such interventions allow us to contemplate improvements in character and behavior of an individual; perhaps in the germ line. We want both the quick cure and the permanent cure without changing our way of life.

This approach misleads us on several scores. It suggests that cure for many diseases may come soon, whereas the success of the gene therapy attempt mentioned above remains doubtful. Furthermore, foreseeable therapies apply only to a small number of single-gene diseases. For the most common disorders such as heart disease, schizophrenia, alcoholism, depression, and cancer, the genetic basis may well be so complex and multigenetic as to render therapy impossible. The costs and risks of these HGP-linked procedures as well as moral and legal considerations should cause us to be cautious in the extreme. Though it has impressed Congress, the HGP's promise of panacea remains uncertain and far from imminent.

Other shadow factors lurk in the background, which should give pause to anyone inclined to speak of our genes as our blueprint and our destiny.* For example, identical (monozygotic) twins have different fingerprints; the rates at which multifactorial diseases strike both twins vary from 17 percent (cancer) to 61 percent (psoriasis). What does this tell us about the power of genes to determine the lives of two individuals whose genomes—46 chromosomes and around 100,000 genes—are exactly the same, as in clones? Since in many cases they share very similar environments, we cannot attribute these differences entirely to that variable. Other forces and factors intervene. As a result, geneticists have borrowed from physics the analogy of a "black box" to represent the uncertainties of gene expression in living persons. This black-box concept conveys the opposite idea from code or fate. Grossly abnormal disorders and behaviors can usually be traced to a single gene. Most inherited conditions are polygenetic and unpredictable by any simple Mendelian table. At least four technical terms have emerged in molecular biology in the effort to identify the uncertainties that limit and modify genetic determinism.† All these considerations and black-box factors should mute the promises the HGP tends to make about curing our individual and social ills. At the genetic level, as at any other physiological, psychological, or

---

*Some information in this paragraph is taken from Vogel and Motulsky.

†*Penetrance* refers to the variable degree to which a gene actually affects or dominates the life and behavior (phenotype) of the individual carrying that gene (Vogel and Motulsky, 84). *Epigenesis* designates the mechanism that, at a certain stage of development, acts on networks of cells to supply variations from any predictable pattern (Changeux, Chapter 7). *Heterosis* causes genes to affect each other in unexpected, nonadditive ways (Konner, 196). *Developmental noise* refers to random molecular events within cells during development, events that produce accumulating deviations from the norm (Lewontin, "The Dream of the Human Genome," 34). All four terms refer to processes as yet little understood.

social level, Goethe's maxim holds: *Individuum est ineffabile.* Individual behavior cannot be predicted or explained.

5. The HGP raises a large number of "ethical, legal, and social implications" so evident and so challenging that the bureaucracy has given them the acronym ELSI. Five percent of the HGP budget is now being devoted to research in that area. In many ways, we will have to confront all over again debates provoked in the first half of the century by the proponents of eugenics. Even without gene therapy, prenatal screening elected by those who can afford it could lead to conditions variously called "hereditary meritocracy," "a biological underclass," "and genetic discrimination." Governments might decide to adopt eugenic policies to improve the national gene pool. Such programs could be highly enlightened—or as sinister as what the Nazis attempted. We think we can now fill the role of Daedalus, the resourceful inventor and artist who served the gods. We are equally likely to play Icarus, his presumptious son, who flew too high for his wings.

An oblique reading on these questions presents itself in the fact that one prominent historian and critic of eugenics, Daniel Kevles, has now become a supporter of the HGP. His career is as revealing of the dynamics of our scientific culture in the nineties as Muller's was for the thirties and Sinsheimer's for the seventies. Kevles' earlier book, *In the Name of Eugenics* (1985), underscored the timeliness of his subject by using a skeptical eye to examine the claims of the great eugenicists. He quoted without approval Galton's statement that "What Nature does blindly, slowly, and ruthlessly, man may do providently, quickly, and kindly" (12). He gave a vivid account of the Fitter Families contests sponsored in the twenties by the American Eugenics Society. Chosen families having the desired physical, mental, and moral traits allowed themselves to be exhibited at state fairs next to livestock displays. The chapter on genetics and molecular biology since World War II is entitled "A New Eugenics"; and the following chapter on the debate about sociobiology was entitled "Varieties of Presumptuousness." The final chapter, "Songs of Deicide," reinforces the impression that we are to read the book as a strong cautionary history in the face of the ambitions of contemporary genetics.

Seven years later, Kevles coedited with Leroy Hood *The Code of Codes* (1992). This collection represents several points of view and contains two essays essentially critical of the HGP. Kevles' contributions, however, do not fall into that category. The skeptical note has disappeared from his voice. His own historical essay closes with a familiar cliché. "The human genome project was steadily gathering the technology, techniques, and experience to obtain the biological grail" (36). I detect no irony. In the concluding essay written with Hood, Kevles must be primarily responsible for the section on the eugenic implications of the HGP. Here he seems to unwrite his earlier book. "It is worth bearing in mind that eugenics was not an aberration, the commitment merely of a few odd-ball scientists and mean-spirited social theorists. . . . Objective, socially unprejudiced knowledge is not ipso facto inconsistent with eugenic goals of some type" (317).

Whatever produced this change of position, Kevles has not responded adequately to the "ethical, legal, and social implications" raised by contributors to his own collection.

The five criticisms of the Human Genome Project presented in the last few pages bring us to one momentous circumstance. It can be seen either as promising or as catastrophic. We can now begin to influence the central process of evolution, natural selection, so as to control it for our own purposes. Without flinching, Kevles quotes Sinsheimer's 1969 prophecy. "For the first time in all time, a living creature understands its origin and can undertake to design its future" (18). Up to now, the "hand" of natural selection guiding evolution has been understood both as invisible, like Adam Smith's free market, and as blind, like chance. Now that invisible hand may gradually turn into our own intrusive hand, bringing direction and purpose where they did not enter before—unless one believes in a divine creator. If the possibility is there waiting for us, could we conceivably abstain like the Princesse de Clèves?

Why *not* intrude? Certain social forces like the welfare state, modern medicine, and birth-control techniques have already interfered with ordinary forces of evolution and have had indirect effects on the genetic composition of the human race. Contemplating that partial and uncoordinated occupation of the territory, some scientists and policy makers contend that we must continue to do so, and with the full benefit of new genetic information and techniques now available.

It does not appear that there is any possibility of our refraining from eating the fruit of that new "tree of knowledge." In such a perspective, the HGP looks like a tiny way station along the road that may lead to our assuming some controls over our own evolution. At the same time, the HGP falls into all four of the categories (practical, prudential, legal, moral) under which I have been raising the possibility of placing limits on scientific endeavor.

4. The Condition of Ambivalence

I have dealt so far in this chapter as much with the claims of and for science as with its deeds and accomplishments. My opening section juxtaposes two arresting statements: one about production of the atomic bomb as "sin"; the other about the Human Genome Project as our "grail." The second section criticizes the claim that we can distinguish pure science from applied science or technology and keep them separate in practice. The five cases I have just examined call attention to the near impossibility of sustaining that distinction under the multifarious pressures of modern life. In their different ways, those cases suggest that there are times when we should consider imposing some limits on scientific activity—most evidently on some of its applications, and possibly on pure research in a few sensitive or dangerous areas. We loathe the very prospect of research to develop chemical and biological warfare and other destructive technologies. Can we simply stop such work? In the case of the human genome, there are as many reasons to slow down and diversify genetic research as to direct it toward an accelerated state-supported project like the HGP.

In the longer perspective, we face the claims of science today much as people in the West have faced the claims of earlier faiths of equal magnitude. Between the ninth and fifteenth centuries, Christianity mobilized all of Europe into a single church, which subsequently broke apart under pressures of reform and doubt that arose primarily from within. The great modern summons to revolution in order to overthrow the oppression of monarchy in favor of res publica soon led to a revulsion from the excesses of revolution. Similarly, the immense appeal of socialism from 1850 to 1980 as a means to improve the quality of life for everyone has also partially collapsed under pressures of reform and doubt from within. Gradually or rapidly, we seem to move toward a condition of ambivalence toward some of our greatest historical constructions—a universal church, egalitarian revolution, the ideal of a socialist society. A comparable and equally deep-seated ambivalence reveals itself today in our attitude toward science, even as science and its applications encroach increasingly on our lives. We seek from science both simple solutions and miraculous transformations of our existence. At the same time, we fear what we believe to be some of the most basic products of scientific inquiry. An examination of a few instances will show how close our hopes lie to our fears—that is, how difficult it is to resolve our ambivalence toward science.

Many myths and legends, including that of Sphinx, tell us that one of our oldest fears is of the monster. A graft of parts from different species, even if accomplished only in imagination, strikes us as an unnatural dissonance in a larger harmony. The Devil himself displays horns, hoofs, and a tail. Frankenstein's monstrous creature anticipates the science fiction of Philip K. Dick (and the movie *Blade Runner*, based on one of his novels), in which laboratory-constructed "androids" are virtually identical with human beings and gradually replacing us. In Dick's story, "The Electric Ant" (1969), the narrator discovers after an accident that he is himself an android. Overwhelmed, he repairs himself for a few hours of intensely gyrating "life" and then commits suicide by cutting his "reality-supply construct tape." He cannot endure his own monstrosity.

Yet the intimate and in some ways monstrous tamperings with our DNA envisioned by the Human Genome Project also present themselves as welcome therapies for cruel diseases. Likewise, the invasive exchange of body parts required for an organ transplant can save our lives. Hybrid strains of certain crops seem to promise huge benefits. New technologies are requiring us to modify our notion of monstrosity along with our notion of what is natural.

Our uncertain feelings about the monster are matched by our response to another scientific product: the chain reaction. Uncontrolled, it suggests pandemic disease or nuclear destruction. But the chain reaction may one day bring us a clean, safe source of energy.

As we might have predicted, an ancient parable representing this double bind has been available for centuries. First recorded in a dialogue by the Greek satirist Lucian and translated into German in the eighteenth century by Wieland, the story was transformed by Goethe into a lilting vernacular ballad. Now, thanks to Walt Disney's *Fantasia* (namely, the segment with Dukas' dramatic score) people all over the world are familiar with "Der Zauberlehrling"—"The Sorcerer's Apprentice." Goethe treats the apprentice's presumptuous unleashing of a broomstick water brigade, a chain reaction, as broad comedy, including the final catastrophe. "Herr und Meister! hör mich rufen!" ("Lord and master! Hear my call!")

Dukas' trumpets and Disney's expressionist animation do full justice to the story. Lucian and Goethe have retold Noah and the Flood without the Lord's preparatory instructions. Instead of a pair of everything being saved in the ark, the Apprentice almost drowns us all together. The outcome seems far more precarious than that of the Bible story. Next time, there may be no Master Wizard to answer our cry for help and to give us one more chance to keep ourselves and our curiosity in check. But could the magic of the chain reaction be harnessed without the Wizard?

The ambivalence we feel toward the external phenomena of science—the monster and the chain reaction—resembles our attitude toward significant inward states: self-knowledge and belief in immortality. From Socrates to Freud, great thinkers have told us that our proper study lies in our own minds and bodies. How could we fear self-knowledge?

I find the bluntest answer in Emerson. In the midst of his succulent and sometimes fatuous prose, he often plants a subtle truth. This one reaches very deep and deserves reflection: "It is very unhappy, but too late to be helped, the discovery we have made that we exist. That discovery is called the Fall of Man" ("Experience").

The mere fact of existence confounds us. But Emerson stops short: Exist as what? I would say that we fear two aspects of the fall into self-knowledge. Science has contributed in a major way to both. One is to discover ourselves alive and locked into some kind of determinism, like androids or puppets or machines. This is the Coppelia complex. In the E. T. A. Hoffmann story "The Sandman," Nathanael finally loses his mind because the dancing doll Coppelia, which he has mistaken for a real flesh-and-blood woman, alerts him to the possibility that he, too, may be a mechanical being, a robot. For Nathanael, the bottom drops out of both reality and identity. In despair, he throws himself from a tower. Molecular biology and sociobiology treat our essential functions as determined and take no account of consciousness and free will. Are we all dancing dolls without knowing it?

The other fearsome aspect of the fall into self-knowledge is to discover that each one of us is alive as a truly free agent, as a unique individual burdened with the responsibility of one's own life. Some areas of neurology and science concerned with the brain emphasize the remarkable differences in development even between identical twins. Such profound uniqueness may exceed our power to sustain it. "God abhors naked singularity," writes Stephen Hawking, paraphrasing Roger Penrose's cosmic censorship hypothesis. One cannot tolerate the thought of oneself as hapax, the single instance of a unique consciousness. We yearn to belong. But without the sustaining presence of any faith or community, many individuals today feel utterly abandoned.

Dostoyevsky's nameless narrator in *Notes from Underground* struggles with equal desperation against *both* these threats of self-knowledge. He fears he is merely a piano key, a cog in a machine housed in a utopian Crystal Palace of scientific determinism. And he fears that his willfully capricious actions in Part Two, designed to liberate his consciousness from all determinisms, have isolated him from every other human being, even from Liza, who appears to understand him. Freed from science, he feels helpless. The Underground Man can find no space to occupy between being no one at all, a mere cog, and being intensely and hysterically himself against everyone else.

Science is also obliging us to ponder again the prospect of immortality. Presumably, most of us seek immortality more than we fear it—seek it at least in one of its many forms: glorious deeds celebrated in history and legend; passing on one's seed to future generations; survival in a lasting material monument; reincarnation; and attaining to spiritual and eternal afterlife. Now we may have to face a sixth fantasy form of immortality: physically living on as oneself indefinitely.

Embracing this line of thought, a few wealthy, optimistic, and self-absorbed individuals unwilling to submit to death have had their cadavers frozen in liquid nitrogen to await the day when science can resuscitate them and keep them alive. Today, a person so inclined could far more simply preserve a DNA sample—as a personal record and to hold open the possibility of cloning a new individual from his genome. Organ transplants represent an attempt at piecemeal rejuvenation. An old philosophical quandary asks how extensively you can darn a sock and still have the original sock. Research on the processes of aging may produce ways of prolonging life. We can no longer chuckle as we once did over stories about the Fountain of Youth. In some form, we may be granted that wish. Most of us would be terrified by the prospect of being incarcerated in one prolonged existence with no foreseeable end. Whether by natural law or ancient custom or both, we conceive human life framed and defined by mortality. Suspension of death appalls us more than death itself. Yet most of us cling to one form or another of immortality, of surmounting mere extinction. Medical science does not reduce, it increases our ambivalence about these final questions of life and death.

We can probably tolerate a good deal of anxiety over the image of ourselves as ancient monsters and over the prospect of knowing how to modify the fundamental givens of human existence. But our ambivalence about traumatic changes in our lives caused by science leads us to consider some constraints on that immense and growing international institution. Events surrounding recombinant DNA tell us that under certain circumstances scientists themselves may restrain their activities in a limited area. But the same case can be interpreted as a false alarm, as a demonstration of how doomsayers can exceed their proper role and of why we should leave science alone. The *Lebensborn* history warns us that scientific knowledge is never safe from exploitation for nonscientific, criminal, and antihuman purposes. It is far less clear how we should evaluate the Human Genome Project.

What, then, can we do about these troubling prospects—the monster, the chain reaction, and immortality itself? Even the simplest and most modest constraint—that of reducing the pace at which researchers and engineers develop new resources and new needs—would be enormously difficult to agree upon and administer. It also flies in the face of hard-earned social, intellectual, and economic principles of freedom. Let me hypothesize. Who will interfere with the radio neurologist who develops a transmittable signal that can hypnotize every person within a two-hundred-mile radius? Our ambivalence reaches very far.

Earlier in this chapter, I raised a question whose time has now come. Is there any existing or hypothetical knowledge whose mere possession must be considered evil *in and of itself*? To any question so purely conceptual, one must answer equally conceptually. As Nicholas Rescher insists, the answer must be: No. Evil and destruction lie only in the mode of acquisition and application of knowledge. But no human life, not even a life dedicated to the pursuit of science, matches the conceptual purity of that question. It takes a situation as contrived as Ulysses sailing by the Sirens to separate knowledge from its acquisition and application. They say the Chinese invented gunpowder and then used it only in firecrackers, not to make firearms. And even then, as gun advocates argue, firearms in and of themselves bring neither good nor evil. The same set of arguments applies to drugs and many other temptations.

But no human moral agent exists apart from the immediate circumstances of a particular life—circumstances that disappear in an abstract question. "Whether of good or evil . . . knowledge cannot defile . . . if the will or conscience be not defiled." Milton's *Areopagitica* makes the best case ever penned for "knowing good by evil." But the "if" clause in the quotation from Milton implies that few human agents are so isolated as to be able to hold knowledge apart from all applications. We must at least be prepared for the worst-case scenario. Those who imposed a temporary moratorium on recombinant DNA research were choosing the prudent course. Even Milton opposed only *prior* censorship; he was not averse to taking action against "mischievous and libellous books" after publication.

While we ponder these questions, an existing precedent is worth contemplating. Most important elected and appointed government officials in a democracy take an oath to uphold the constitution under which they serve. Accordingly, they can and should be held to a higher standard of conduct and responsibility than the ordinary citizen. By ancient tradition, physicians subscribe to an oath that acknowledges both the great power of their professional knowledge and their responsibility to use it judiciously and cautiously. The wording of the Hippocratic oath, only spottily administered today by medical schools in the United States and throughout the world, lacks grace but carries immense weight.

> *You do solemnly swear, each man by whatever he holds most sacred, that you will be loyal to the profession of medicine and just and generous to its members; that you will lead your lives and practice your art in uprightness and honor; that into whatsoever house you shall enter, it shall be for the good of the sick to the utmost of your power, you holding yourselves far aloof from wrong, from corruption, from the tempting of others to vice; that you will exercise your art solely for the cure of your patients and will give no drug, perform no operation for a criminal purpose, even if solicited, far less suggest it; that whatsoever you shall see and hear of the lives of men which is not fitting to be spoken, you will keep inviolably secret. These things do you swear.*

Such a ritual oath cannot change human nature. It defines a profession and alerts everyone, within and outside the profession, to its principles and ideals and to possible abuses. And such an oath calls upon the profession to exercise some restraints on its members if they do not observe it. The fact that many doctors today are either unfamiliar with the oath or ignore it does not detract from its symbolic importance. Precisely this form of allegiance to a set of principles consecrating a body of knowledge and governing its practice constitutes the kind of responsible tradition our culture needs in the professions. Medical schools would do well to revive it, and medical societies to discuss its meaning today.

Scientists have produced no comparable oath, no such symbolic recognition of special powers entailing corresponding duties. Possibly the fields of science are too numerous and dispersed to bring together under a single statement. Nevertheless, everything I have said in this chapter convinces me that the effort to draft such a statement would be worthwhile. To make this proposal, I have had to overcome a deep-seated revulsion from anything that resembles the loyalty oaths required of college professors by some states in the 1950s and 1960s. But humanity as a whole, rather than an individual state, has a legitimate interest in the professional loyalty of a scientist to whom we have given long and privileged training.

A life devoted to science has become as much a stewardship as holy orders or knighthood or public office or medicine.

It is rewarding to find an eminent scientist and physician sympathetic to my assessment. Late in *Physician to the Gene Pool,* James Neel, while discussing "humankind's genetic dilemma," confronts similar questions about the responsibility of scientists. He introduces four major recommendations with a section entitled "*Primum non nocere.*" "This Hippocratic aphorism, enunciated for the guidance of physicians some 2000 years ago, remains as valid today as then: *Above all, do no harm.* . . . Although Hippocrates wrote for physicians, the aphorism is equally relevant for geneticists" (344–45).

If they were to take an oath including the injunction "*primum non nocere,*" those receiving doctorates in scientific fields would be encouraged to assess scrupulously the consequences of their work, to study pertinent cases in the history of science, to avoid the cooption of their work by inappropriate agencies, and to produce a more principled profession. Most scientists already have a fairly well-developed professional conscience. An effort to define and declare that conscience might at the present juncture help scientists scrutinize the proliferation of research in dubious areas and renew the confidence of ordinary citizens in a profession that appears to have inherited the mantle not merely of evolution but of revolution.

But let us not become too solemn about these matters and too inclined to see science as a menace. C. S. Peirce, the great American philosopher of science, came to believe that we should not conceive of science as a body of systematized and applied knowledge. It constitutes "a mode of life" devoted not to the truth as one sees it but to "the truth that the [scientist] is not yet able to see but is striving to obtain" (*Values in a Universe of Chance,* 268). In *Science and Human Values* (1956), Jacob Bronowski insists that the tradition of science properly practiced generates sound values comparable to those previously based on faith and authority. Science relies on "the habit of simple truth to experience [which] has been the mover of civilization."

Science is neither a sin nor a grail. Not our child but our invention, science as a discipline will never grow up to think for itself and to take responsibility for itself. Only individuals can do those things. We are all the stewards of science, some more than others.

The knowledge that our many sciences discover is not forbidden in and of itself. But the human agents who pursue that knowledge have never been able to stand apart from or control or prevent its application to our lives. Despite the tale of Odysseus and the Sirens, "pure research" is a modern myth. Therefore as science explodes in a few areas into a vast enterprise propelled as much by commerce and the waging of war as by curiosity, we need to scrutinize this disproportionate growth. The free market may not be the best guide for the development of knowledge. State planning has not always served us better. While we ponder these wrenching questions, let us not forget the stories of Icarus and of Bacon's "Sphinx," and the very different case histories of Himmler's *Lebensborn* program and the Manhattan Project. In this era of liberation and permissiveness, it may well be that a judicious oath for scientists will help to prevent us from acting like the Sorcerer's Apprentice.

Chapter vii

THE DIVINE MARQUIS

■

No girl was ever seduced by a book.

—Jimmy Walker, Mayor of New York

Forty years ago, before television established its dominance, many civil libertarians ridiculed Fredric Wertham, the senior psychiatrist at Bellevue Hospital in New York City, for his warnings about the serious effects of comic-book violence on the behavior of children and adolescents. The basest form of profit motive unredeemed by social responsibility drives the publishers of comics to depict escalating mayhem, Wertham wrote in *The Seduction of the Innocent* (1954). Their product does not have the dreamlike quality of fairy stories to temper and deflect the violence. More than twenty years later, without referring to comic books and television, Bruno Bettelheim's *The Uses of Enchantment* (1976) argued eloquently that fairy tales teach children not to imitate cruelty and destructiveness but to overcome them. Bettelheim was defending an essential part of our cultural heritage and, by implication, deploring the effects of the media.

Meanwhile, a related debate about our moral environment has addressed obscenity (a legal term for a category of materials that does not qualify for First Amendment protection) and pornography (a literary term for works intended primarily for sexual arousal). Following the liberating sweep of the 1960s, a series of major government reports and court decisions during the 1970s, in Europe and the United States, removed virtually all restrictions on obscene and pornographic materials except for their distribution to minors. The U.S. Supreme Court decision in *Miller* v. *California* (1973) has stood for more than twenty years and makes prosecution of obscenity close to futile.

One of the immediate effects of these developments was to render both possible and highly profitable the translation and publication of works by an author who represents the extreme case of forbidden writing. For almost two centuries, the Marquis de Sade had lain buried and preserved in the cultural unconscious of Europe. Furthermore, his republication now gave impetus to a twentieth-century move to rehabilitate Sade and to present him as a great revolutionary author. He even benefited from a curious double presumption in his favor: He had spent time in prison; his works had been censored. Do we need any further proof of his heroic stature? But the Sade revival reaches deeper into our literary and moral thinking than this special case of "affirmative action" for the persecuted. We half-believe that the rediscovery and release of repressed experience will heal a split in our being and free us to live more fully. And we tend to misconstrue the Freudian phrase "beyond the pleasure principle" by projecting it onto a shadowy realm of cruelty and destruction that merits exploration. The Sade revival feeds both these contemporary tendencies.

Pornography we shall have always with us. It serves a purpose and in its traditional forms poses no serious threat to decency and morals. The healthiest reaction is usually laughter, not outrage. But the life and works of the Marquis de Sade raise particular problems. Powerful claims have been made about his importance as a moralist, philosopher, and novelist. His writings represent the most challenging test case of a forbidden author, a case I cannot avoid, given the subject of this book.

1. THE SADE CASE

During the first two weeks of July 1789, crowds in Paris began to assemble around the Bastille. They believed that there must be people imprisoned in the royal fortress with whom they sympathized and whom they might liberate from the king's despotic rule. The crowds did not know that only a few aristocrats were held there, mostly on morals charges, and scarcely deserving liberation in the name of the people. On July second, the crowd in the street heard a voice, apparently amplified by a megaphone improvised out of a rain spout, shouting that prisoners were being slaughtered and needed rescue. The prisoner to whom that voice belonged was summarily transferred two days later to the Charenton insane asylum. When the crowd stormed the Bastille on July fourteenth, it ransacked his cell and pillaged or destroyed his personal belongings, his furniture, his six-hundred-volume library, and a considerable collection of manuscripts, some carefully hidden in the wall.

The man willing to attempt so outrageous a ruse in order to regain freedom after twelve years of imprisonment we know today as the Marquis de Sade. After his death, his name provided the word *sadism.* Following a series of scandalous incidents, he had been arrested in 1777, primarily due to the relentless pursuit of his mother-in-law. Police and court documents have recorded the serious morals charges brought against Sade in four principal incidents in Paris, Arcueil, Marseille, and at his own Château de La Coste near Marseille. They included homosexual and heterosexual sodomy (both capital offenses at that period), various whippings and possible knifings of prostitutes, masturbating on a crucifix, corruption of young girls, death threats, and other "excesses." As a result of these charges Sade was burned in effigy, imprisoned several times by order of the king, shot at by an incensed father, and condemned by the high court of Provence to decapitation. While his persevering wife tried to help him, his mother-in-law would not forgive him for absconding to Italy with her other daughter. A lettre de cachet obtained by the mother-in-law finally precipitated Sade's arrest in 1777 in Paris, where he had traveled on the occasion of his own mother's death.

From prison, Sade immediately protested, that "My blood is too hot to bear such terrible harm," and he threatened to commit suicide. A few years later, he wrote to his wife that the revolting abstinence imposed on him "brought my brain to the boiling point. It causes me to conjure up fanciful creatures which I shall bring into being." He was bragging as much as he was complaining. After a short stint in solitary confinement in the fortress at Vincennes, Sade was allowed books, candles, writing materials, and exercise periods. His reading included Cervantes, Rousseau, Voltaire, Mme de La Fayette, Prévost, Marivaux, Laclos, Richardson, Boccaccio, and classics he had read earlier at the rigorous Jesuit school Louis-le-Grand. In the 1780s, Sade began marathon sessions of writing, which continued when he was transferred to the Bastille. During a thirty-seven-day burst, using both sides of a forty-foot roll of paper, he composed the uncompleted draft of his encyclopedic opus on debauchery and crime, *The 120 Days of Sodom*. When that manuscript disappeared after the storming of the Bastille, Sade despaired over the loss of his earliest and most ambitious work. Later, he turned over to his wife two enormous manuscripts: an epistolary-picaresque novel of love, corruption, and travel, *Aline and Valcourt*; and at least two versions of *Justine, or Good Conduct Well Chastised*, a work so licentious and sexually explicit even for those times that Sade never acknowledged it as his.

This thirty-seven-year-old nobleman and cavalry officer was descended from two ancient and distinguished aristocratic families in Provence.* Sade was brought up close to the royal household in Paris in an atmosphere of licensed debauchery. One of his fictional letters is often quoted as shrewd self-analysis: "[This childhood] made me naughty, tyrannical, irascible; it seemed to me that everything should give way to me, that the whole world should condone my caprices, and that it was up to me alone to plan and satisfy them." In spite of his high connections and several properties in the South of France, Sade was forever in financial straits. He had married for money at twenty-three. His wife bore him two sons and a daughter, while he frequented brothels and spent wildly on actresses.

When the Constituent Assembly abolished the royal lettres de cachet in 1790, Sade recovered his liberty and lost his wife, who

---

*One Avignon tradition held that Petrarch's Laura was married to a member of the Sade family.

obtained a legal separation. He tried his hand at writing plays and published both *Aline and Valcourt* (signed) and *Justine* (anonymous). He also began living with Constance Quesnet, an impecunious actress who had been deserted by her husband and had a son. She remained his companion for the next twenty-four years until his death, a period during which Sade attained a certain notoriety but neither wealth nor security.

During the giddy days of regicide and the approaching Reign of Terror, Sade's activities in the revolutionary Section des Piques in Paris and as a pamphleteer counterbalanced his vulnerable status as an aristocrat opposed to the Terror. Arrested in 1793 as an enemy of the Revolution, a false patriot, and a libertine, he barely escaped the guillotine; he was released in 1794. The next several years brought severe hardship. He published four more licentious books, including *Philosophy in the Boudoir* and *Juliette*.

Sade's third major imprisonment, which lasted the last thirteen years of his life was occasioned not by his politics or his personal morals but by his writings. With Napoleon now in power, the prefect of Paris and the minister of police imprisoned Sade in 1801, without the scandal of a public trial, as the author of "that infamous novel, *Justine*, and the still more terrible, *Juliette*." Sade spent most of these years in the Charenton insane asylum, where his obesity became pronounced. For a considerable time, he was allowed to produce plays for the inmates as a kind of therapy. The valor of his son in battle encouraged him to petition for his freedom directly to Napoleon. No response came, and he remained in Charenton at his family's choice and expense. Coulmier, the director of Charenton, generally supported Sade in opposing harsh plans to transfer him to a prison, as proposed by the chief medical officer and by an inspector. They considered him a criminal, not a lunatic. Napoleon himself in privy council decided against moving Sade to a prison. Until the end, he continued to corrupt sexually young girls and boys living in the asylum. He died at age seventy-four of pulmonary congestion and a "gangrenous fever."

One can argue that in the eighteenth century, political events and philosophical thought in Europe and America performed the double feat of liberating us from the domination of priests and kings and of creating a civil society based on law and representative democracy. In France, the human cost was terrifying. The many heroes of that century include figures as varied as Voltaire, almost a king; Rousseau, almost a prophet; Tom Paine, almost a hero; and Thomas Jefferson, almost a statesman. Their eminence resides in the historic events they precipitated and the institutions they helped to form.

What, then, can we do with the Marquis de Sade, who belongs to the same era? Should we try to fit him into that current of liberation and new political institutions? Hardly. He remained a haughty aristocrat, despised the Revolution, and imagined his own moral and philosophical Reign of Terror that would solve all his problems and rid him of his "persecutors." He emerges as a thorn in the side of the Enlightenment, a man who carried revolutionary libertinism to patently undemocratic extremes of argument, narrative imagination, and personal behavior. It is very difficult to separate the notoriety of his person, the tocsinlike insistence of his ideas about God, nature, and man, and the sheer scandal of his writings. Today we attempt to deal with these aspects separately as biography, philosophy, and pornography. In his career, they overlap and interfere with one another, particularly when he defends himself. From prison, he wrote clamoring letters to his wife.

> *I am guilty therefore of pure and simple libertinism, the like of which is practised by all men, more or less due to their varying temperament or inclination.*
>
> (1781)

> *Not my manner of thinking but the manner of thinking of others has been the source of my unhappiness. My fanaticism is the product of persecutions I have endured from my tyrants.*
>
> (1783)

Such protests of innocence to a person he trusted suggest that Sade had deluded himself about the cause of his troubles. Yet in his fictional writings when he sought dramatic effect, he did not hesitate to describe as "turpitude" and "depravity" the behavior he calls "pure and simple libertinism" in the passage above.

There may be a physical source for Sade's obsessions. Recent biographers, especially Maurice Lever, document the fact that in spite of a "volcanic" sexual temperament, Sade had difficulty achieving orgasm. Two practices aided him: pain, inflicted and undergone, and passive sodomy, simulated in prison with special dildos ordered for him by his wife. How pertinent are such considerations? It begins to appear that Sade's sexual excesses, real and imagined, may arise as much from some pathological impediment as from hot-bloodedness.*

Another uncertainty hangs over Sade's case. What lies at the heart of his project? Why did he devote his time in prison to writing when he knew the manuscripts might well be seized or lost? Had he found a mission, or was he merely killing time by indulging his fantasies to the full?

We may find answers to these questions in the way Sade addresses his reader, *and the censor*, at the beginning and the end of his books. He took two approaches. In *Justine* (1791) for example, he describes the crescendo of sexual ordeals undergone by that hapless beauty. Yet Justine survives with her physical health and her innocence intact—only to be blasted at the end for her virtue by an act of God, a lightning bolt that violates her succulent body once and for all. Virtue cannot win. Then Sade carefully frames the story so as to serve pious ends. This is no licentious tale, he maintains. The rhapsodic dedication to his "enlightened" companion, Constance, insists that he composed the sequences of debauchery and torture "with the sole object of obtaining from all this one of the sublimest parables ever penned for human edification." Both "sublimest" and "edification" are ambiguous enough under ordinary circumstances to provoke a guffaw. But we are not reading Rabelais. On the closing page of the same book, the bolt of lightning converts Justine's depraved sister, Juliette, into a devout Carmelite nun. Returning to his initial unctuousness, the

---

* A coded 1784 letter to his wife from the Bastille referring to "La Vanille and la Manille" apparently concerns his difficulties in masturbation. Justine is saved from her first ravisher, Dubourg, by "the loss of his powers before the sacrifice could occur." Sade was at some pains to describe his liaison with Constance Quesnet as nonsexual. After citing different evidence, the scholar Raymond Giraud concludes: "To me it seems impossible to mistake the only slightly indirect confession of sexual inadequacy." Simone de Beauvoir suggests that Sade was in part impotent.

author-narrator urges the reader to be convinced, like her, of "the same moral. . . . that true happiness is to be found nowhere but in Virtue's womb." (We learn in the sequel that the conversion does not last.)

In the preface to *The Crimes of Love* (1800), Sade constructs an impressive historical essay on the novel form and insists on the originality of his own fiction in showing men modified by vice and passion. "Never, I repeat, never will I depict crime as being anything but the work of hell." A paragraph on instructing men and correcting their morals opens *Eugénie de Franval,* the story of a rich nobleman who carefully rears his daughter to become his slavish mistress at fourteen. Three times in three pages in the prefatory "Notice" to *Aline and Valcourt* (1793), Sade cites "the moral purpose" of "painting vice" in order to make people detest it.

In the narrative portions of this group of novels, Sade's fictional characters produce passionately reasoned speeches in defense of immorality, crime, selfishness, torture, and generalized destruction as they accomplish such actions. In the didactic passages that frame these narratives, Sade took pains to establish their status as negative object lessons and to redirect the stream of smoking debauchery back into the channels of morality. He found the same argument as Milton in *Areopagitica* and, indirectly, in *Paradise Lost*: namely, that ignorance of vice and of its temptations permits only "blank virtue" and a "puppet Adam." Is Sade sincere in these claims, as some critics and readers seem to believe? Or does he address the reader here, with a wink and a leer toward the censor?

In another set of works, Sade sings a different song. In these cases, he could forget the censor because he did not intend publication under his own name. A one-page dedication "To libertines" opens *Philosophy in the Boudoir* (1795). "Voluptuaries of all ages, of every sex, it is to you only that I offer this work; nourish yourselves upon its principles: they favor your passions. . . ." Sade's earliest extended work, *The 120 Days of Sodom,* nowhere makes a claim to be morally edifying. Its one hundred pages of preliminaries insist in trumpet tones recalling Rousseau's *Confessions* that the greatness and originality of his work reside in its being "the most impure narrative ever written since the world began." Is this his true stance?

Anyone who has read more than a sampling of Sade's writings

will, I believe, feel dissatisfied by such a reduction of the discussion to two alternatives: Sade favors depravity; Sade opposes depravity. We are dealing not with an ordinary mind but with an already dissolute one forced by prolonged imprisonment to confront daily the fascination of its own proliferating phantasms and unable to put them aside for the ordinary activities of a free man. Unlike the Princesse de Clèves and Emily Dickinson's poetic persona, who tempered their ardent impulses according to the revelations of an utterly lucid imagination, Sade exploited his imagination to stoke the furnace of his lusts. The elaborate constructs of his fictional orgies and the running hyperbole of his style leave the impression of a man engaged in a self-imposed task, a man writing on a bet. He has wagered with himself and with his persecutors that he can and will systematically invert every human virtue—above all, Christian virtue. His repertory of evil will compose so unique a project as to be completely original and utterly scandalous to his enemies. He will work his revenge and construct his monument in a single enterprise. His writings look like the result of the resolve of an obsessed monarch to build himself a tomb that will command attention forever to the greatness of his excesses. Sade's personal excesses were directed primarily toward sodomy and corruption of the young. His monument celebrates every sexual crime imaginable—literally. Here lies the meaning of his "fanaticism" in adhering to his principles and tastes, the fanaticism defiantly affirmed in his 1783 letter to his wife. Sade's literary bet grows into a project of monumental defiance. I have discovered the word *gageure* ("wager") only once: in a wager, the Duc de Blangis (*The 120 Days*, 27) has himself sodomized fifty-five times in one day. Later in the book, the duke makes, and always wins, a few more bets *(paris)*. But the sense of a wager—against God, against humanity, against civilization itself—drives all Sade's heroic characters and the author himself even more obsessively than Faust's wager drives him.

After Sade's death his works did not disappear; they went underground. His reputation was kept alive by virulent attacks and by journalistic clucking sounds acknowledging the prurient appeal. More canny, Sainte-Beuve remarked in 1843 that Byron and Sade were having an enormous influence on modern writers, "the latter being clandestine, but not too clandestine." Flaubert and Baudelaire left brief, vivid statements on Sade in their letters and notebooks. Symbolists and decadents in France and a few writers like Swinburne in England encountered Sade's writing without disturbing his clandestine status. But the saga of the Marquis de Sade had not come to an end. For the tide would turn.

2. Rehabilitating a Prophet

The strong man with the dagger is followed by the weak man with the sponge.

—Lord Acton

In 1810, the Bibliothèque Nationale began receiving a "legal deposit" of every book published in France. In order to accommodate licentious and obscene works and restrict their readership, the library created a special collection, which soon received the name of *Enfer* or Hell. (American libraries sometimes used the Greek letters Δ or Z.) These works formed a kind of official underground; to consult them, one had to receive special permission. Sade's works and manuscripts were placed in this well-secured institutionalized Hell as well as in many private collections. During the nineteenth century, a few copies circulated under the counter at high prices.

Around 1850, the great French historian Michelet appeared to have found where to classify Sade in the order of things: as the ultimate representative of a corrupt monarchy and "professor emeritus of crime." "Societies end with these kinds of monsters: the Middle Ages with a Gilles de Rais, the celebrated child killer; the Ancien Régime with a Sade, the apostle of murderers" *(History of the French Revolution)*. The attention of the medical profession to Sade's life and writings began to modify that outcast status toward the end of the nineteenth century. Krafft-Ebing's *Psychopathia sexualis* (1886) coined the word *masochism* and appropriated *sadism*, which had already entered French in the 1830s. It reached English later. By 1901, in a work called *The Marquis de Sade and His Works Seen by Medical Science and Modern Literature*, Dr. Jacobus warned against the dire effect of reading his "bloody" novels. It was a

Berlin doctor, Iwan Bloch, who in 1904 unearthed the long-lost manuscript of *The 120 Days of Sodom* and published the first limited and faulty edition as a work related to Krafft-Ebing's case histories. The scene was now set for Sade's rehabilitation, a tangled intellectual story never fully told. I divide it into four stages.

The first stage covers the first four decades of this century. During the effervescent decade in Paris before World War I, the young poet Guillaume Apollinaire was recognized as a leader of the avant-garde. His mastery of free and formal verse, his astonishing, if erratic, erudition, his Rabelaisian imagination, and his close association with young artists soon to be named the Cubists—this rare combination of talents did not earn him an adequate living. His exotic tastes led him to the Enfer of the Bibliothèque Nationale, where he prepared a series of licentious works published under his editorship. The most important title in the collection, *The Work of the Marquis de Sade: Selections*, appeared in 1909 with Apollinaire's fifty-page introduction commandingly entitled "The Divine Marquis." The fairly reliable biography and summary of major writings contain a series of ringing statements about Sade's scientific contribution to the psychopathology of sex and about his neglected cultural significance.

> *The Marquis de Sade, the freest spirit that has ever lived, had particular ideas on women and wanted them to be as free as men. . . . One of the most astonishing men that has ever appeared. . . . This man, who seems to have counted for nothing during the whole nineteenth century, might become the dominant figure of the 20th.**

Just before he died at the age of thirty-eight, Apollinaire met a young man, Maurice Heine, with whom he made plans to publish Sade's works. Heine introduced André Breton and the Surrealists to Sade's writings and in 1929 traveled to Berlin to rescue the manuscript of *The 120 Days of Sodom* from the German doctor who had botched the first edition. Heine set about an ambitious program of

*Apollinaire welcomed hoaxes and mystifications and concocted one himself by passing off as journalism a totally fictitious account of Walt Whitman's funeral. In his hyperbolic introduction to Sade, commercial motives and prankishness play a role. But Apollinaire had also done extensive research to support a seriously argued literary opinion.

Sade publication and research. A reliable version of *The 120 Days of Sodom* appeared in three volumes between 1931 and 1937. The first Surrealist Manifesto (1924) mentioned Sade as "surrealist in sadism." In a lecture delivered at Oxford in 1936, the Communist-Surrealist poet Paul Eluard called Sade "more lucid and pure than any other man of his time" *(L'évidence poétique)*. In 1933, the Italian critic Mario Praz published the original version of *The Romantic Agony*, an influential book on decadence, which gives careful and critical attention to Sade. The following year, a young British anthropologist, Geoffrey Gorer, produced the first of three editions (also 1953 and 1962) of *The Life and Ideas of the Marquis de Sade*. This early version emphasized the philosophical and political connections between Sade's ideas and Nazism. Later editions assert that Sade had a scientific contribution to make to the pathology of sex. In a brief essay entitled "Beliefs," collected in *Ends and Means* (1937), Aldous Huxley devoted two pages to Sade in which he refers to "the philosophy of meaninglessness carried to its logical conclusion" and ends with a claim echoing Apollinaire and Eluard. "De Sade is the one complete and thorough-going revolutionary of history."

Known mostly by reputation, difficult of access, and surrounded by an aura of dangerous seductiveness, Sade had the status in this period of a rare archaeological site with an ancient curse to protect it. Early Sade scholars resembled explorers of an exotic outpost of the human record. It is not surprising that they made extravagant claims about this genuinely extreme case. They did not know then that Sade's excessive writings and their excessive estimates of him would fuse into a powerfully infectious intellectual strain. When, a few decades later, the results of these explorations became commercially profitable, there was no stopping the influx.

During the 1930s, two young French authors were taking Sade very seriously and very personally. The neo-Nietzschean philosopher Pierre Klossowski studied Sade's "liquidation of the notion of evil" and his restoration of it in the notion of crime as a form of forbidden knowledge—"*crime-connaissance.*" Georges Bataille expressed outrage at the Surrealists' appropriation of Sade, for they had no conception of his truly excremental vision and shirked the duty not just to imagine his excesses but to *practice* them. (Bataille had plans for trying out human sacrifice.)

This intermittent attention to Sade over a thirty-year period, from Apollinaire to 1939, constituted a rumbling, not yet a revival. The wind really shifted in the decade after World War II. An important critical work or a new edition appeared almost every year. In this second stage, Sade was released from confinement in Enfer and began to be accepted by some publishers and readers alongside standard literary authors. The list of critics who accomplished this rehabilitation include some of the most illustrious names of the period. One could speculate that the gradually revealed horrors that had been taking place in Germany and the Soviet Union drove some readers to seek shelter in the imaginary and seemingly harmless horrors of Sade. But his rehabilitation remains difficult to explain. I attribute it more to an eerie post-Nietzschean death wish in the twentieth century. That death wish seeks absolute liberation, knowing that it will lead to absolute destruction—physical, moral, and spiritual. For some, apocalypse exerts a strong attraction.

The most influential and frequently reedited essay of the sequence is the earliest—Jean Paulhan's forty-page "The Marquis de Sade and His Accomplice," written in 1946 as the introduction to the second edition of *The Misfortunes of Virtue,* and translated in the Grove edition of *Justine.* It opens chattily to reveal "the secret" of the New Testament Gospels: that "Jesus Christ is light of heart . . . he enjoys himself." This preliminary discussion plants the idea of looking for a new Gospel and for a secret or mystery in Sade. Paulhan grants briefly that criminals are dangerous and must be punished. But Sade, in Paulhan's version, was barely a criminal and, anyway, he "paid, and paid dearly" with thirty years in prison. Furthermore, criminals in general are more interesting than law-abiding people. "I mean more unusual, giving more food for thought." Therefore, Sade deserves sympathetic examination, especially since, according to Paulhan, there is no more cruelty and violence in his writings than in Las Casas' *The Destruction of the Indies* (1575). The consequences are evident. Preoccupied by "looking for the sublime in the infamous," our best contemporary literature is "dominated, determined by Sade as eighteenth-century tragedy was by Racine." In a rhapsodic sentence filling a page, Paulhan affirms Sade's "unfaltering demand for the truth" and concludes with a shorter declaration: "His books remind one of the sacred books of the great religions."

It would be difficult to surpass that claim, but Paulhan has much more to say. He produces an anodyne definition of sadism that reduces its tortures and murders to a selfish version of utilitarianism. "We demand to be happy; we also demand that others be rather less happy than we are." The closing pages of the essay return to the first motif of a secret, a mystery.

> *Sadism, in the final analysis, is probably nothing else than the approach to and, as it were, the (perhaps maladroit and certainly odious) testing of a truth so difficult and so mysterious that once it is acknowledged as such . . .* [it becomes] *instantly and miraculously transparent.*
>
> (34)

Behind all the labored disclaimers in that sentence, what *truth* is Paulhan talking about? Why the sibylline tone instead of explicit statement? I surmise that Paulhan thinks Sade revealed the awful secret that our supreme pleasure can be achieved only through pain—pain to ourselves and, more commonly, pain inflicted on others. But the basic question is not who discovered the phenomenon but how widely it applies to human beings. Paulhan implies without evidence that sadism is a universal truth and he presents to us an alluring and essentially innocent Sade, untainted either by the real crimes he committed himself or by the obsessive outrages against humanity he imagined and welcomed in thousands of pages of writing.*

The year after the appearance of Paulhan's essay, his friend

*Paulhan's writings of the period call for a halt to the prosecution of Nazi collaborators. His reasoning has earned a certain notoriety. The French Resistance was organized in great part by Communists who, before the war, had sought to overthrow the French Republic and to collaborate with the Soviet Union. Those now accused of collaboration with the Germans had served as resistance leaders in the earlier struggle against Moscow. So where was the moral high ground? Everyone—except for de Gaulle and company and non-Communist *résistants* like Paulhan himself—had sold out to one totalitarian enemy or the other. Paulhan's call for evenhandedness and amnesty based on the long view of history leads to a moral helplessness, or abdication, that also characterizes the Sade essay. We cannot condemn war criminals when others may have behaved as badly, Paulhan argues. Accordingly, we cannot condemn Sade, whose writings have made us all his accomplices. To understand is to forgive.

Georges Bataille published in his own review, *Critique*, an article on Sade that later became the centerpiece of his book *Literature and Evil* (1957). First a sympathizer with Surrealism, then its opponent, Bataille wrote several pornographic novels, founded the Collège de Sociologie (with Roger Caillois and Michel Leiris), and developed a philosophy based on a belief in sacred destruction and excess—a mystical nihilism close to Nietzsche's cult of Dionysus. Bataille's thinking and writing carried him toward erotic violence and human sacrifice and toward the disorder and destruction he saw in Hitler and Stalin.

Today, Bataille may be best known in intellectual circles for having proposed an elaborate theory of transgression. Some find great profundity in it; others, an immature yet sinister perverseness. For a few privileged individuals (a limitation usually left unstated), prohibitions on crime and violence exist not so much to inhibit action as to add zest and intensity to those actions experienced as transgression. Bataille uses Hegel's dialectical term—*aufheben*—to make his case. "Transgression should not be confused with a reversion to 'the state of nature'; it suspends *[lève]* the prohibition without suppressing it . . . a complicity between the law and breaking the law" (*L'érotisme*, 39). Bataille finds the ultimate embodiment of transgression in Sade's writings. By denying any form of solidarity with other people, transgression expresses itself as "impersonal egoism"—a strikingly accurate term for both Sade's and Bataille's outlook.*

---

*The introduction to *Eroticism* contains some unblinking declarations."To violate is the secret of eroticism. On any scale, eroticism is the domain of violence, of violation." In the notes, Bataille explains that he means both physical and moral violence. "Eroticism is born of interdiction, it lives on interdiction" (695). These seemingly analytical statements carry a programmatic message. If *all* erotic behavior is by definition transgressive and violent, then there is no such thing as perverted or pathological eroticism, just as there is no normal, nonviolent, loving eroticism. Like Kinsey (*Sexual Behavior in the Human Male*, 1948, which Bataille had read with care), Bataille has attempted to demonstrate that no sexual behavior is deviant and everything is permitted. But he salvages the feeling of sinfulness in order to maintain the excitement of naughtiness. Furthermore, Bataille suffers from a strong physical revulsion in the domain of sexuality. "The body is a thing; it is vile." He links erotic activity not to reproduction or to pleasure but to pain and death.

Bataille's shocking treatment of his wife and daughter can be documented in Marcel Moré, *G. Bataille et la mort de Laure*.

The most reliable author of our day on sadomasochist behavior, the medical doctor and psychoanalyst Robert J. Stoller, wrote extensively and understandingly

A century earlier, Baudelaire gave lyric expression to transgression in large segments of *The Flowers of Evil.* The title expresses metaphorically what emerges unadorned in the last line of "The Unremediable":

Tête à tête sombre et limpide
Qu'un coeur devenu son miroir!
Puits de Vérité, clair et noir,
Où tremble une étoile livide,

Un phare ironique, infernal,
Flambeau des grâces sataniques,
Soulagement et gloire uniques,
—La conscience dans le Mal!

("*L'Irrémédiable*")

Dark confrontation of the heart!
Once it becomes his [the Devil's] looking-glass,
A well of Truth, a black morass
Where one pale trembling counterpart,

Dancing firebrand of the Devil,
Comfort and glory of the mind,
Shines like a lighthouse for the blind
—Conscience wallowing in Evil!

(TR. WALTER MARTIN)

In his book on Baudelaire, Sartre reproaches the poet for holding on to a sense of sin and evil as dramatic lighting for the sordidness of his life. Sartre presents this attitude as a classic case of "bad faith." When Bataille reviewed Sartre's book on Baudelaire, he could not escape the evidence. "Sartre is right: Baudelaire chose to be guilty,

---

about violence and violation in S&M haunts in San Francisco. After twenty years of such research, Stoller did not lose his perspective. The opening pages of *Pain and Passion* (1991) insist that "the desire to harm others" is not the secret or the principle of all eroticism. Hostility and hurt in sex represent "aberrant behavior" and a "perversion"—a word that Stoller insisted on maintaining.

like a child." Bataille wishes to be guilty also: His own theory of transgression is childish, a refurbished secular form of sin to keep life interesting for select intellectuals with kinky tastes.

In a 1947 essay entitled "Sade," a key to much of his later writing, Bataille quotes Swinburne, Apollinaire, and Paulhan in support of Sade's genius and of the significance and beauty of his literary works. Bataille approaches the crux of his argument in a long passage on Sade's "moral situation."

> *Very different from his heroes in that he showed human feelings, Sade experienced states of frenzy and ecstasy which seemed to him in many ways possible for all. He judged that he could not or should not eliminate from life these dangerous states, to which insurmountable desire led him. Instead of forgetting them in his normal state, as is the custom, he dared to look them right in the eye and faced the enormous challenge they pose to all men. . . . In the solitude of prison Sade gave the first reasoned expression to these uncontrollable movements, on whose negation consciousness has founded the social edifice—and the image of man.*
>
> (*LA LITTÉRATURE ET LE MAL*, 141)

Here is a defense brief for the worst criminal conduct that might lurk behind Mr. Hyde's "undignified pleasures." Behind the claims about Sade's daring and originality, Bataille has introduced with the phrases "insurmountable desire" and "uncontrollable movements" a theory of fate or determinism tinged with admiration—the idea that Sade and his fictional heroes should not be held responsible for their actions. The next paragraph describes how the frenzied excesses of *The 120 Days of Sodom* soil, blaspheme, and demolish everything in sight. But for Bataille, Sade has grasped the truth.

> *In reality, this book is the only one in which the mind of man measures up to* what is. *The language of* 120 Days of Sodom *is that of a universe in slow and sure decline, which tortures and destroys the totality of beings to which it gave life.*
>
> (143)

"*What is,*" then, means the total destruction of everything, a dream that rivals God's power to create everything. Bataille also quotes Paulhan on "seeing the sublime in the infamous, and greatness in the subversive." Bataille would have us conclude that Sade's most depraved scenes of systematic perversion, torture, and murder constitute a new sublime—like the wild beauty of violent storms and terrifying precipices in nature: elevating spectacles. The editors of Bataille's writings in English translation refer to them as "liberating."

A few years later when Albert Camus was writing a study of rebellion and murder in the contemporary world, he found he had to assign a major place to Sade. In *The Rebel* (1951), Camus does not share the enthusiasm of Paulhan and Bataille for Sade's works. He concentrates on the life. Camus' fourteen carefully written pages on Sade acknowledge him as the first "metaphysical rebel"—against God, against man, against everything. He rebels in the name of absolute freedom, which, according to Sade, warrants all crimes, including murder. The discussion of Sade, the first and longest devoted to any individual figure in *The Rebel,* presents him as a negative example. For Camus, Sade got everything wrong.

> *Sade's success in our day is explained by the dream that he had in common with contemporary thought: the demand for total freedom and dehumanization coldly planned by the intelligence. . . . Two centuries ahead of time and on a reduced scale, Sade extolled totalitarian societies in the name of unbridled freedom.*
>
> (46–47)

By the end of *The Rebel,* we understand the meaning of Bataille's "in reality." Camus wants no part of Sade's absolutism and nihilism. "If rebellion could found any philosophy, it would be a philosophy of limits." Camus' sobering paradoxes of freedom express profound annoyance with intellectual fashions that promote Sade, yet a puzzling indifference to writings that have inspired those fashions.

Simone de Beauvoir's eighty-page tract "Must We Burn Sade?" (1952) nowhere mentions Camus. But its timing associates it with the great public joust between Camus and Sartre over the criticisms of Stalinism in *The Rebel.* Half-rebutting Camus on Sade, Beauvoir belittles the biographical figure and fastens on the writings. "Sade's eroticism doesn't lead to murder but to literature." Unimpressed by Sade's "original intuition" that coitus and cruelty are identical, she begins with a blanket judgment. "Neither in his life nor in his work does he surmount the contradictions of solipsism."

But Sade has worked his wiles on Beauvoir, and her argument falters.

> *Thanks to his stubborn sincerity, even though he is neither a skillful artist nor a coherent philosopher, Sade deserves to be acknowledged as a great moralist.**
>
> . . .
>
> [Sade] *approves of vendetta, not of courts of justice; one can kill but not judge.*
>
> . . .
>
> *The immense merit of Sade is to proclaim the truth of man against any escape mechanism of abstraction and alienation.*

By the end, the solipsist has become a realist "attached to the concrete world." Beauvoir discerns well enough that Sade's "truth" amounts to egoism bordering on madness. But she has not decided how to respond to his challenge and concludes lamely that "The supreme value of his testimony is that it troubles us."†

This second, decidedly mixed stage of Sade's rehabilitation coincided with the relaxation of censorship and obscenity standards in France, the United Kingdom, and the United States. The trials during the 1950s and 1960s concerning *Fanny Hill, Lady Chatterley's Lover,* and (in Paris) four of Sade's most licentious books opened the way for legal publication and distribution of all Sade's works in inexpensive editions. Financial profit added new excitement to the celebration. Grove Press in the United States invested heavily in a long-term project to translate Sade.

---

*In French, *moraliste* means not so much moralizing person as author of reflections on human customs and on the human condition, like Montaigne and Pascal.

†We must not overlook the melodramatic circumstances of the Camus-Sartre controversy during which the essay was written. I am convinced that for Beauvoir Sade becomes the symbol or analogue of Stalinism. She perceives the horror of both but is not yet ready to denounce or reject them.

The third stage of Sade's rehabilitation began in the 1960s. It moved increasingly toward vindication and became a torrent of writings. In his preface to the *The Misfortunes of Virtue* in 1965, Gilbert Lely speaks of the transformation of Sade's works from "monstrous and criminal" into "masterpieces of French literature." The same year *Yale French Studies* published a special number on Sade, and Klossowski, one of the first in the field, added a revisionist lecture to his earlier articles to form a probing yet cryptic volume, *Sade mon prochain* ("Sade my neighbor," 1967). Two new biographies in English appeared a few years later, and many women joined the fray on Sade's side in the 1970s and 1980s.

One of the most significant essays on Sade, first published in German in 1944 and not translated into English until 1972, slipped sideways into the debate at about this time. The two major founders of the Frankfort school of sociology, Max Horkheimer and Theodore Adorno, include in *The Dialectic of Enlightenment* a forty-page excursus entitled "Juliette or Enlightenment and Morality." Their prose is by no means easy to follow, and they make the serious error of classifying Sade as a bourgeois rather than an aristocratic figure. But like Klossowski, they are wrestling with a demon. The argument is tantalizing.

They maintain that anyone who accepts Kant's attempt to derive morality from reason is a "superstitious fool." Only Sade discovers a moral order based on the self-preservation of the bourgeois individual. In their elaborate organization "without any substantive goal," Sade's orgies should be seen as a form of "modern sport," the free play of reason. They offer us "the pleasure of attacking civilization with its own weapons"—that is, with systematic rational planning. In the elegiac and murky closing pages, the authors seem to be saying that Sade shows us the true face of Enlightenment reason—cruelty as greatness, totalitarian-state socialism, and the Homeric epic of domination. It is unclear whether we are to admire this prospect or to take fright at it. Horkheimer's and Adorno's ambivalence toward Sade corresponds to the ambivalence of the entire book toward the collapsed project of Enlightenment reason.

The other major figure of this period of the Sade revival shows no such uncertainty about how we are to interpret Sade's historic role. Michel Foucault presents as fundamental for the emergence of the modern era out of seventeenth-century classicism the fact that Sade revealed to us the truth about man's relation to nature. Foucault plants his declarations at crucial junctures in his two major works of 1961 and 1966. These four passages reveal the usually obscured center of his ethos.

> *Sadism . . . is a massive cultural fact that appeared precisely at the end of the eighteenth century and that constitutes one of the greatest conversions of the occidental imagination . . . madness of desire, the insane delight of love and death in the limitless presumption of appetite.*
>
> (*Madness and Civilization*, 210)

> *Through Sade and Goya, the Western world received the possibility of transcending its reason in violence. . . .*
>
> (*Madness and Civilization*, 285)

> *After Sade, violence, life and death, desire, and sexuality will extend, below the level of representation, an immense expanse of darkness, which we are now attempting to recover . . . in our discourse, in our freedom, in our thought.*
>
> (*The Order of Things*, 211)

> *Among the mutations that have affected the knowledge of things . . . only one, which began a century and a half ago . . . has allowed the figure of man to appear.*
>
> (*The Order of Things*, 386)

The last quotation from the final page of *The Order of Things* does not allude to Sade by name. But, in association with the other passages and in context, there can be little doubt that the great cultural "mutation" welcomed by Foucault refers directly to Sade's moral philosophy and to its actual practice in life. "A Preface to Transgression," Foucault's 1963 essay, appropriates Bataille's paradoxes in order to glorify Sade's transgressive language. The incantatory writing of the essay veers often into incoherence and mystification. We should be fully aware of Foucault's outlook behind the flamboyant style. For him, Sade is our saviour.

During this third stage, few writers warned *against* the rush to revive Sade as our new prophet and savior. But before the libertarian seething of the 1960s had reached full pitch, one well-informed book appeared that for the first time gave unblinking attention to Sade as a thinker, placing him in the context of eighteenth-century intellectual history. In Lester G. Crocker's *Nature and Culture: Ethical Thought in the French Enlightenment* (1963), Sade is accepted into the big league of philosophical thinkers with Hobbes, Hume, Voltaire, and Diderot. After the sections on "Natural Law," "Moral Sense Theories," and "The Utilitarian Synthesis," Sade occupies Crocker's stage nearly alone for forty pages on "The Nihilist Dissolution." Sade's ideas are cited throughout the book. On the other hand, Crocker denies to Sade any originality in ideas or literary form. As evidence, he offers us a series of quotations from La Mettrie's *Le bonheur* (1748), a book Sade read with care.

> *In regard to felicity, right and wrong are quite indifferent . . . a person who gets greater satisfaction from doing wrong will be happier than anyone who gets less from doing good . . . there is a special kind of happiness which can be found in vice, and in crime itself.*
>
> . . .
>
> *Let pollution and orgasm make your soul, if it is possible, as sticky and lascivious as your body . . . I urge only to peace of mind in crime.*

Sade could find in La Mettrie's ideas the equivalent in philosophy of the Terror in political history. D'Argens and L'Abbé Dulaurans, also familiar to Sade, wrote novels almost as licentious and egoistic as his. But Crocker affirms that Sade was the first to construct a "complete system of nihilism with all its implications, ramifications, and consequences. . . . Nihilism is the worm at the core of our culture. It is the flaw we must constantly overcome." Crocker's thesis owes much to Camus. The synopsis and quotation of Sade's ideas on nature and culture gradually describe a philosophical system as scandalously inhuman as the criminal acts enacted in the novels.

Crocker even appears to have found the elusive "truth" so many have pointed to in Sade.

> *Justice is the supreme folly, for it bids us to attend to the interests of others, not to our own. In actual practice, we find "just" whatever is in our interest, according as we are weak or strong. . . . And why, Sade demands, should we hide from ourselves such a truth?*
>
> (410–11)

Crocker fails to mention how many of these arguments were made by Callicles—and answered by Socrates—in Plato's *Gorgias*. Crocker skillfully points out the contradictions in Sade's nihilistic synthesis that annul it as a tenable philosophical position. For Crocker, Sade was "the first to face the failure of rationalism" and "foretold the course of the crisis of Western civilization." No other discussion of Sade's ideas in their historical context, not even in French, rivals Crocker's for concision and clarity. Unfortunately, he accepts the hackneyed opinion that Sade's "sexology" has a contribution to make to modern psychology.

Crocker was a leading scholar of eighteenth-century French letters and philosophy. Roland Barthes, one of the influential *maîtres à penser* of the 1960s and 1970s, held the Chair of Semiotics at the Collège de France, a position created especially for him. In publishing *Sade, Fourier, Loyola* (1971), Barthes seized upon Sade as a subject for opposite reasons and with opposite effect. Crocker chose Sade to occupy the nihilist niche in his analysis of ethical thought in Enlightenment France and thus assigned him a place in the history of Western philosophy. Barthes needed Sade (flanked by Fourier and Loyola) to provide the final test and proof of his semiotic system, which interprets all writing as belonging to a pleasurable edifice of signs unrelated to reality and beyond moral judgment. I believe that Sade would not have been pleased by Barthes' conversion of his works into "texts," into a harmless set of linguistic structures and grammatical devices, a mere combination of words with no impact on our lives. More accurately, Sade would have been incensed by Barthes' cavalier defanging of his life's work.

Barthes' preface goes right to work to obliterate Horace's classic notion of literary works as *dulce et utile*—"both pleasing and instructive." In Barthes' quiet revolution, the first word simply usurps the second.

> *The pleasure of reading something guarantees its truth. By reading texts rather than works, by approaching them so as to seek not their "content" or their philosophy but their delight in writing, I can hope to remove Sade, Fourier, and Loyola from their usual sponsorship* [caution]*; I try to disperse or avoid the moral discourse that has grown up around each of them.*

Barthes says that he will "steal" them away from the all-pervading bourgeois ideology in order to listen to the violence of their "excess . . . as writing."

After the mannerisms and analytic tricks of his earlier critical work, *S/Z,* this triple study displays Barthes' full resourcefulness as a reader. He has spotted genuine correlations of organization and style among Sade, Fourier, and Loyola. Barthes' allocation of pages and the outline of chapters suggests that his greatest concern is to rescue Sade from any censorious judgment. After a masterful inventory of Sade's "protocols"—hidden strongholds, food and dress, money, character portraits—he proceeds to treat Sade's most horrible episodes as mere permutations on erotic postures perfectly comparable to a language using words and grammar to form sentences. Everything reduces to a set of codes, and "the sentence . . . converts the network of crime into a marvellous tree [of erotic ramifications]." One example Barthes gives of these ingenious language constructions fits into a single succinct sentence out of Sade. "In order to unite incest, adultery, sodomy, and sacrilege, he buggered his married daughter with the host." Barthes wants it to sound as ordinary as algebra or a crossword puzzle.

Barthes concludes that Sade writes true "poetry," a displacement of ordinary language by pure "writing." He contends that since Sade engages in no kind of mimesis or imitation and practices no realism, since the depraved and criminal Juliette is made merely "of paper" or of words, and is "not frightening because inconceivable in reality," we have no reason to condemn or censor these writings. In his view, they merely tell stories and play brilliantly imaginative word games that give us much pleasure. We might be playing Scrabble.

But Barthes could not fully defang Sade's writings by placing them in a refurbished art for art's sake tradition. Having read Barthes' writing on Sade as well as that of others, the Italian poet and radical Communist, Pier Paolo Pasolini adapted a Sade novel for what turned out to be his last and most scandalous film, *Salò, or The 120 Days of Sodom* (1975). The garish publicity surrounding the film's release mingled with the revulsion provoked by the brutal street murder that ended Pasolini's bruised life as an avowed cruising homosexual. Possibly the most deliberately outrageous mainstream film ever produced, *Salò* depicts cruel sexual perversions so rigorously enforced that the one rebellious moment of ordinary heterosexual love is immediately punished by execution of both parties. It is not a comic scene. These ghoulish episodes of coprophilia, sodomy, and torture explicitly enacted are linked by the setting to fascism in its last throes. Pasolini's many statements imply that he saw the unrelenting obscenity and violence of the film as an attack on "the Power"—that is, the commodification of everything, including sex. *Salò* can also be seen as an extreme instance of that very power in the form of calculated provocation and scandal.

An Asian author and a European director collaborated to revive Sade in a less explicit and more subtle fashion than did Pasolini's film. The Japanese writer Yukio Mishima, who committed hari-kari in 1970, published *Madame de Sade* in 1965. The circling dialogue among six women keeps the Marquis de Sade offstage while his devoted wife repeatedly refers to him as golden-haired, tender, sweet, and loving. Then at the end of the play (which omits the fact that she dropped him utterly when he was released from prison), she collects into a revealing coda most of the Promethean and Faustian pronouncements made about Sade in the twentieth century. "He is the freest man in the world. . . . He piles evil on evil, and mounts on top. A little more effort will allow his fingers to touch eternity. . . . [He has] created holiness from the filth he has gathered."

Staged in Swedish by Ingmar Bergman in the 1990s, the play earned international acclaim as a theatrical masterpiece. The Sade rehabilitation had proceeded so successfully that no critic I read was prepared to question the premises and the purport of the play. The drama critic of *The New Yorker*, John Lahr, called the performance a "noble" evening and then turned his attention to monsieur, not to madame. "Perversion became an act not of debasement but of discovery . . . evil itself becomes a miracle."

After Pasolini, Mishima, and Bergman had thrown the doors open even wider, celebration of Sade's depraved universe became almost commonplace. One frenzied celebrant, Camille Paglia, does not hesitate in *Sexual Personae* (1990) to quote utterly explicit passages from Sade that illustrate the link between sexual pleasure and acts of torture and murder. On this basis, she sees Sade as "a great writer and philosopher whose absence from university curricula illustrates the timidity and hypocrisy of the liberal humanities." But Paglia also relies on the same alibi of aesthetic distance as does Barthes.* Here, art entails no responsibilities; it escapes judgment. "Literature's endless murders and disasters are there for contemplation, not moral lesson." After a particularly horrible quotation, she writes: "Remember, these are ideas, not acts." Or she emphasizes Sade's "comic gratuitousness." But somehow Sade comes to count for a great deal. In followers like Paglia, he has found a cult.

The fourth stage of Sade's rehabilitation constitutes final consecration. We have now encountered claims that Sade was the freest of all revolutionaries, the inventor of a new sublime, a great moralist of transgression, and a poetic word artist with no moral dimension. His consecration as a standard author among the masters took place twice, first in 1989 in the pages of *A New History of French Literature*, edited by Denis Hollier of Yale University and published by the Harvard University Press. In this revisionist history, three hundred five-page entries focus successively on a precise date at which a literary event took place—often the publication of a major work. In the seventeenth century, for example, Corneille, Molière, and Racine each qualify for one full entry, Saint-Simon for none. Out of nine centuries of writing, only one author merits *two* full entries in this book: the Marquis de Sade.

Summer 1791 is chosen for inclusion because of the appearance of *Justine*, anonymously. After comparing *Justine* to Voltaire's *Can-*

---

*The fact that Paglia savagely attacks Barthes and Foucault on many scores does not prevent her from accompanying them on the Sade rehabilitation project.

dide, Chantal Thomas speaks of Sade's novel as "a particularly admirable example of this dynamic of terror" and links it to popular *histoires tragiques* "written to restore the reader's virtue through the spectacle of vice." "Passions in Sade, though undoubtedly excessive and systematically transgressive, are never implausible." Apparently, Thomas had neglected to read Barthes. She also has her version of truth in Sade.

> *The libertine confronts an ineluctable truth, that of the absolute egoism of pleasure. . . . Thanks to this indifference, which is precisely what does not exist in what is commonly called Sadism, a principle of detachment, a lightness, underlies Sade's writing.*
>
> (583)

Let us not be misled. The sangfroid Sade advocates in carrying out the most bestial tortures consists in the systematic elimination of all feeling for other people, in favor of infantile egoistic pleasure. The cold-bloodedness of Sade's sadism resembles sheer depravity, whose heavy hand displays no "lightness"—least of all in any sense connoted by Milan Kundera's novel *The Unbearable Lightness of Being.* Thomas concludes: "Contrary to all received opinion, Sade's name should in fact evoke the image of innocence victimized." One learns to scrutinize carefully any sentence containing the words *in fact.*

The 1791 entry in the Harvard *New History* carries the title "Pleasure, Perversion, Danger." The March 1931 entry, entitled "Sadology," picks up the story with Maurice Heine's corrected edition of *The 120 Days of Sodom.* Carolyn J. Dean, the author of these pages, traces the rediscovery of Sade in the twentieth century and begins by quoting Apollinaire's prophecy that Sade would become its "dominant figure." Dean does not pause to question. "[Apollinaire] was right. . . . Sade would now be praised as an unhypocritical—albeit extreme—expression of nature, testifying to the variety and complexity of natural impulses rather than to an individual pathological depravity." Without a murmur, the sentence swallows as true Sade's running claims about the naturalness of all the crimes he advocated and described. A later phrase of Dean's refers almost in passing to "the horrifying, inassimilable core of his experience, linking sexuality not to pleasure but to terror." Precisely. Like most of those associated with Sade's rehabilitation, Dean welcomes this association of sex with selfishness, cruelty, crime, and murder.

After these encomiums, Dean's last sentence tries to hoist itself back into the world of realities. "During the 1930s the surrealists' celebration of Sade began to appear terribly naïve in the face of a kind of political violence that seemed to replicate some of Sade's horrible scenarios" (894). This hesitant demurrer follows two pages vigorously arguing the opposite case. *Naïve* is a singularly meek word to apply to a dreadful misjudgment. Furthermore, the stricture applies far less to the Surrealists than to later champions such as Bataille and Blanchot, whose writings justify this entry on "Sadology."

Both the 1791 and the 1931 entries in *A New History of French Literature* nod briefly toward the sheer horror of Sade's writings and then belittle it through the use of a trivializing concessive. Thomas writes: ". . . though undoubtedly excessive and systematically transgressive." Dean's version is more succinct: "—albeit extreme—." Thus they relegate the center, Sade's inhuman excesses, to the periphery and talk about him in terms of acceptable abstract concepts such as nature, transgression, irony, creativity, and "the triumph of desire over objective reality." This new history attempts to present Sade as a dominant and admirable figure in French literature.

Sade's second consecration consists in his publication in 1990 in the respected and handsomely printed French series, the Bibliothèque de la Pléiade. The honor corresponds to an artist's work being admitted to the Louvre. Michel Delon's lengthy introduction, which justifies Sade's selection for the series, traces the partial disappearance and later reappearance of his writings. It also lays out for those who read carefully an argument in favor of Sade that sometimes prevaricates and sometimes tells all. Sade's works, Delon says at the start, have rallied around his cause all those opposed to the institution of literature. Romantics, decadents, esthetes, Surrealists, the Tel Quel group—all these writers "have encouraged *[conforté]* their refusal of the bourgeois order and of moral dogmatism by reading [Sade's] novels that obliterate good conscience." All of this, it goes without saying, is a good thing in Delon's eyes. At the end, he invokes Michel Foucault in favor of Sade and implies that Foucault's evaluation of Sade is based on aesthetic appreciation, as one might admire a string quartet. Delon does not mention Foucault's ringing endorsement of Sade's moral nihilism. This omission is partly restored by Delon's own view of how aesthetics usurps morality. "If the beautiful is content to imitate nature as codified by the Ancients, the sublime must strike us in the manner of savage and undisciplined nature, the nature that unleashes storms and volcanoes, the nature that drives pyromaniacs and torturers" (Sade, Pléiade, LVI). After this defense of the sublimity of crime exactly echoing Sade and Bataille, Delon's closing sentences exult over the restitution of Sade to great literature. "Sade inaugurates in literature an era of suspicion toward every power and toward every discourse. . . . Without banalisation, without provocation, Sade belongs in the Bibliothèque de la Pléiade."

Is it possible that the India paper, limp leather binding, and scholarly apparatus of the Pléiade edition can transform Sade into an author to be read along with Dickens, Balzac, and Melville with pleasure and profit by our own children? "The strong man with the dagger is followed by the weak man with the sponge." Does Lord Acton's quip about history apply to literature? Almost all the literary and philosophical discussions of Sade I have mentioned sponge away the depravity and the bloodiness of his narratives by considering only his ideas. The Pléiade edition will let it all show, will withhold no horrors. But our task as readers has only begun. For after the rehabilitation and the double consecration, we will do well to look carefully in order to find out whether, inside the handsome binding, the dagger is still at the ready and what it is pointed at.*

---

*Since Sade's consecration, large numbers of books and articles have continued to appear that seek to reinforce his status as a "canonical" author. In many cases, it is difficult to decide whether the critic is profoundly naïve about human nature, or disingenuous, or both. Peter Cryle's *Geometry in the Boudoir,* for example, wishes to emphasize literary tradition. The book examines "classical erotic literature" as a genre according to Gadamer, Sade's narrative techniques of counting, geometrizing, and modeling, and the whole vexed question of "canon formation." Nowhere, in a study intensely aware of Sade's sustained seduction of the reader through the appeals of writing, does Cryle try to reckon with Sade's *message,* with what his teachings mean to our lives. Cryle deals confidently with form and simply ignores the challenge of Sade's content. Or so it appears. Through such approaches, Sade is now being taught in a number of colleges and universities.

In order to approach this task, I treat two specific cases of the probable influence of Sade's ideas and narrative scenes on an exceptional personality and then go on to a direct consideration of Sade's works. It makes a long way around. But I believe that the detour serves to open our eyes to the full significance of his case.

3. The Moors Murders Case

The couple's name was Smith. Their call at 6:00 A.M. on October 7, 1965, to the Staleybridge police station in Hyde, near Manchester, England, was answered by Constable Antrobus. We are not now entering the burlesque world of Ionesco or Thornton Wilder, who use precisely these names in their plays. David Smith, a seventeen-year-old husband with a police record for violence, had decided to report to the police what he had been lured by a close friend to witness the previous night: a gory ax murder. Smith had accompanied Myra, his wife's sister, back to her apartment just before midnight and had walked in on a scene that had been staged in part for his benefit by Myra's boyfriend and Smith's own buddy Ian Brady. For months, Brady had been initiating first Myra and then Smith into a universe of crime and murder through lengthy discussions, books of sadomasochist exploits, and Nazi propaganda. That evening, Brady killed a homosexual youth in front of Smith and tried to get him to participate. After helping to clean up, Smith went home. A few hours later, he and his wife, both terrified, decided to report the crime and thus turn in their closest friend and a family member. Later, during the enormous publicity of the trial, at which he was star witness, newspapers paid Smith substantial sums of money for interviews and exclusive stories. The defense tried hard to implicate Smith in the murder he had witnessed and reported.

Thus began the Moors murder trial of 1966 in the beautiful medieval walled town of Chester. Given massive coverage in the English newspapers, it appalled the public for three weeks. The previous months' investigations had unearthed two earlier murders of even younger victims, whose bodies had been buried on the nearby Saddleworth moor. For my account, I have relied on the almost daily articles in The *Times* (London) from April 19 through May 17, 1966, as well as on subsequent books.

Less than two years earlier, in November 1963, Brady and Myra Hindley had lured John Kilbride, fourteen, to their apartment and subjected him to "sexual interference, shortly before he was killed," according to the medical examiner. Brady photographed Hindley standing on the fresh grave on the moors. Two years later, after the murder of the homosexual boy, they abducted Leslie Ann Downey, ten, on Boxing Day (December 26) from a fair. In Hindley's apartment, they gagged and stripped the child and tape-recorded her pleas and screams while Brady took pornographic photographs of her and killed her—probably by smothering. The photographs, shown to the jury, are not described in newspaper accounts. The photography expert, who testified that they were taken with Brady's camera, could say only that "no adjective in the English language is appropriate" for these pictures. When the sixteen-minute live tape recording of Leslie Ann Downey was played from the well of the courtroom in total silence, the jurors heard footsteps, muffled voices, some clear pleading such as "Please, God help me. . . . Please, Mum. . . . What are you going to do with me? . . . I want to see my mummy," whimpering, and then horrible screams. Two women in the public gallery covered their ears. Other people in the courtroom contained their feelings until they heard the Christmas bells and music Brady and Hindley had dubbed in at the end: "Jolly Old Saint Nicholas." At the opening of the trial, a defense motion had been granted to remove all women from the jury and replace them with men.

Four times during the trial, the attorney general, acting as prosecutor, referred to the collection of fifty books found in a suitcase, including *Orgies of Torture and Brutality*, *History of Torture through the Ages*, *Sexual Anomalies*, *Cradle of Erotica*, Geoffrey Gorer's *The Life and Ideas of the Marquis de Sade*, and Hitler's *Mein Kampf*. On April twentieth he read out of "an orange colored book" Sade's defense of murder as "necessary, never criminal." Smith identified the passage as one read aloud to him by Brady, who admired Sade's works and praised him as "a good author." The next day, the defense cross-examined Smith about his indoctrination by Brady and read passages out of Smith's notebook. "Mr. Hooson asked Smith to turn to page 24 and read: 'Rape is not a crime, it is a state of mind. Murder is a hobby, and a supreme pleasure.' Smith replied, 'That is not my mind. This is I surmise the mind of the Marquis de Sade' " (April 23).

Brady later acknowledged his ownership of the books in the suitcase. On May second, the prosecution returned to the Evans murder, reported by Smith, which opened the whole case. Brady is being cross-examined.

> THE ATTORNEY GENERAL: *But there was nothing more sordid than killing this boy? —Yes. But it depends on how you think.*
>
> *You don't go with the view of the Marquis de Sade on these views on murder, do you? —I have read de Sade and it's Smith's book.*
>
> *You have read it and enjoyed it. —Yes.*
>
> *And approved of it? —Some of it.*
>
> *And the bits about murder? —No.*
>
> *The Attorney General then referred Brady to a list of books which he said he did not want to read out. He asked Brady: "They are all squalid pornographic books?"*
>
> BRADY: *They cannot be called pornographic. They can be bought at any bookstall.*
>
> (MAY 3)

Throughout the trial, all the participants maintained a meticulous courtesy except Brady and Hindley, who were alternately stubborn and defiant. Their lawyers tried hard to implicate Smith and to taint his evidence. Apparently, the tape, the photographs, and the other evidence convinced the jury. In two hours and fourteen minutes, it convicted Brady on three murder counts and Hindley on two. Because of recent legislation abolishing capital punishment, they escaped hanging and are still serving multiple life sentences.

The Moors trial decided the question of guilt on the charge of murder. It did not try to establish any conclusions about two matters raised at several points and left unresolved: the extent and nature of sexual molestation that accompanied the murders and the degree of influence on the accused of the books named and quoted in court. Smith, under intense pressure from Brady to open his mind to murder and other crimes, testified that he vomited when he returned home after witnessing the murder of Evans. Asked why a few hours later he called the police and reported his best friends, Smith answered: "I could never have lived with myself." That ultimate cliché also contains a succinct description of what used to be called the voice of one's conscience. Socrates in the *Euthyphro* and the *Apology* says he relied constantly on a "divine sign." The intended corruption of Smith, far from a model character or citizen to begin with, came up against a limit, human or divine, that Brady and Hindley had left far behind.

Halfway through the trial, a columnist in *The New Statesman* measured the newspaper coverage of the event during the first week. *The Express* led with 690 column inches—about four full columns (not tabloid size) per day, starting on the front page. The *Times* held itself to half that. Editorial writers worried about the harm such uncontrolled reporting might have on the public. Having attended the trial, Pamela Hansford Johnson published a probing book the following year. *On Iniquity* (1967) opens with a careful recapitulation of the murders as reconstructed from courtroom testimony. Johnson reflects on the moral questions raised by the events, particularly on the relations between affectlessness in many lives and the stimulus to the imagination of obscene and pornographic writings. At the extreme point of this deeply troubled book, Johnson wonders if "cruelty, like crime, is imitative." In *Beyond Belief*, published the same year, the playwright Emlyn Williams composed a semifictionalized narrative of Brady's early life and crimes. Williams' conjectural scenes, though plausible, carry less weight than his documenting of Brady's steady consumption of violent crime films and of books on Nazism and on or by Sade.

As Johnson predicted, when the Moors murders disappeared from the headlines, they seemed to fall almost totally out of public consciousness. Newspapers in the United States gave comparatively little space to the trial. They would have the opportunity before many years passed to deal with an even more ghastly case.

4. TED BUNDY'S SERMON

In January 1989, after ten years of legal maneuvering and last-minute stays, the state of Florida prepared to execute by electrocution a convicted serial killer aged forty-two. Ted Bundy's all-American appearance and fluent speech seemed to magnify the horror of his abductions, rapes, bludgeonings, and strangulations. Here was a role model gone astray. When the NBC nightly news reported the early-morning execution and the ghoulish revelries that had greeted it in Starke, Florida (placards read: BURN, BUNDY BURN; ROAST IN PEACE), the program included clips of an interview with Bundy taped the previous afternoon. Viewers were encouraged to stay tuned for the full interview a few minutes later on *Inside Edition*. Watching it, some people had the feeling that Bundy had survived his own death, that the monster had come back to haunt us through an electronic afterlife. Or was the Bundy interview the modern equivalent of holding up on a pike for the bloodthirsty crowd Bundy's severed and dripping head? In any case, for half an hour he answered at length brief and leading questions from Dr. James Dobson, a Christian evangelist from Pomona, California, president of Focus on the Family, and a trained psychologist.

Born illegitimate, the young man sometimes called the "Preppie Killer" had a highly stressful childhood in Philadelphia in the house of a tyrannical grandfather. When he was four, his mother moved with him to Washington State, married, and had a family. Bundy grew up in a modest, suburban, churchgoing household in Tacoma. He did all right in school, excelled at psychology at the University of Washington, and failed at law school. He had learned about being illegitimate and had been rebuffed by one beautiful upper-class girl. Bundy worked for the Seattle Crime Prevention Advisory Committee and during two campaigns impressed Republicans with his talents and his promise. Meanwhile, unknown to anyone, Bundy had been abducting, sexually abusing, mutilating, and then throttling young women with long dark hair, usually students. During the 1970s, at least forty victims in four states fell into the hands of this attractive confidence man, perhaps over one hundred. Arrested and convicted in Utah in 1977 for aggravated kidnapping, he later escaped in the middle of his trial in Aspen to Florida, the state fellow prisoners told him had the strictest capital punishment statute for murder.

Compared to the clueless perfection of his earlier crimes, there was something willfully public about the way he smashed the skulls of four Florida State University students in their sorority house, killing two. Three weeks later, he picked on a twelve-year-old. After his arrest in 1978, the Florida justice system found itself dealing with a confident, even contemptuous criminal whose famous facc and horrible exploits inspired several books and a TV movie. In two successive trials, the jury found him sane and guilty of first-degree murder. Then came ten years of legal maneuvering to avoid the electric chair, which he appeared to have sought out by escaping to Florida. When all avenues of appeal and delay had failed, Bundy turned contrite and confessed to a series of unsolved crimes, promising many more if granted more time. The governor of Florida, Bob Martinez, caught the situation just right: "For him to be negotiating for his life over the bodies of victims is despicable."

Bundy himself had initiated a correspondence with Dobson in 1987 and requested the last-minute interview. Here is how two national magazines summed it up.

> *The condemned man linked his crimes to violent pornography and alcohol. Bundy said that as a child he had become fascinated by sexual violence that "brings out a hatred that is just too terrible to describe." He said that alcohol reduced his inhibitions against killing.*
>
> (*MacLean's*, February 6, 1989)

> *He said that he had been brought up in a good Christian home. He said that he had been a normal person, but that violence in the media—pornographic violence in particular—had brought out in him a hatred "that is just, just too terrible." Other criminals, he said, had been similarly driven and maddened by pornography. Bundy paused at times and spoke of other things, but his interlocutor brought him back to his theme.*
>
> (*The New Yorker*, "The Talk of the Town," February 27, 1989)

Both accounts are accurate and incomplete. A column by John Lee in *U.S. News & World Report* distorts the truth. "Bundy offered the traditional left-wing explanation for crime ('the environment made me do it'), but with a traditional right-wing twist (the environment was pornography, not poverty or discrimination)" (February 6, 1989).

Here is what Bundy said: "I'm not blaming pornography; I'm not saying that it caused me to go out and do certain things. And I take full responsibility for whatever I've done. . . . The question and the issue is how this kind of literature contributed and helped mold and shape the kinds of violent behavior."

Later he came back to the point. "I don't want to infer that I was some helpless kind of victim. And yet, we're talking about an influence. . . ."

The editoralists at *The New Yorker* suggested a plausible, unsentimental response to the interview, one probably adopted by a large segment of the magazine's readers.

> *It was not, in truth, a very interesting exchange. As is often the case with television interviews, there were few surprises in it; no voices were raised or doubts encouraged. After one glance at the setup, you understood its premises and prearrangements; you could have turned off the set at any point and not missed much . . . that silent grinding is the medium's mammoth forgettery.*
>
> (FEBRUARY 27, 1989)

Not an interesting exchange? We shall have to examine that proposition. *The New Yorker* appeared intent on having us dismiss the interview as prepackaged.

Anyone who has explored Bundy's record with any thoroughness will find other cogent reasons to be skeptical about the interview. This pariah who had come to feed on publicity found a way to manipulate the media into staging his apotheosis on one of the three major networks during prime time. His first expression of remorse after ten years of stonewalling amounted to a confession—the only newsworthy element of the entire exercise. Second, the staged event gave Bundy an opportunity to raise the pornography issue and thus to displace some of the guilt for his crimes onto society, even as he appeared to accept responsibility for his acts. Third, every available record shows Bundy as a practiced, systematic liar with no appeal to truth, only to expedience and concealment. Why should we believe anything he said on any subject?

Fourth, the format of a star fed easy questions by a compliant interviewer after adequate rehearsal offers the least likely method of investigating a subject and of searching out the truth. Lastly, the very sponsorship of the show—Dobson's Focus on the Family, which plays an active role in opposing pornography—may discredit its content as a foregone conclusion.

Could there be any reasons to listen to what Bundy said? Whatever further celebrity and personal satisfaction Bundy might have achieved by his last-minute confession, the remarks about pornography could gain him nothing. In the muted form in which he made them, they seem unsensational, perhaps therefore "uninteresting." Second, we are very unsure about who should be considered an expert on the question of the connection between violent pornography and crime. Considering his own unspeakable crimes and his ten years in prison and on death row with other felons, Bundy cannot be dismissed as lacking pertinent experience. Third, the statements about pornography in the Dobson interview are consistent with what Bundy had said before many times without causing any protest. The authors of the most dispassionate book on Bundy, Stephen Michaud and Hugh Aynesworth, two men who interviewed him in relays over a period of several years, make this point clearly. "[We were] amazed by the brouhaha once Dr. Dobson's tape is released. There is nothing in it that hadn't been in print since 1983, save for Ted's twist on the devil-made-me-do-it defense"* (*The Only Living Witness*, 353).

Thus, Bundy repeated in the final interview observations not invented for the occasion or produced for Dobson's benefit alone. It is possible that he spoke from experience. It is possible that the essentially commonsense, even clichéd, things he said about pornography deserve our attention. We cannot deny him the status of a privileged witness, so long as we remain properly skeptical. What, then, did Bundy say?

Dobson opened the interview with the high drama of bare fact. "You are to be executed tomorrow morning at seven o'clock." Then, very soon, the crux. "Where did it start?" Bundy described his "fine Christian home." At twelve, he discovered soft-core drug-

---

*"The devil-made-me-do-it defense" refers not to the effects of pornography but to Bundy's theories about his split personality.

store pornography, followed by more explicit books. "The most dangerous are those that involve violence and sexual violence." The craving for excitement those books fed went through stages like those of an addiction "until you reach the point where the pornography only goes so far. You reach that jump-off point where you begin to wonder if maybe actually doing it will give you that which is beyond just reading about it or looking at it." His inhibitions gradually weakened. Alcohol helped. Here, Bundy made the point that not everyone is affected this way: ". . . some people would say that, well, I've seen that stuff, and it doesn't do anything to me."

After discussing Bundy's background, Dobson wanted to know about his crimes. Twice he asked, "What was the emotional effect on you?"—once in general, once in respect to the last gruesome sex murder of twelve-year-old Kimberly Leach. Bundy refused the latter question. "I won't be able to talk about that. I can't begin to understand." To the first question, he answered with a description of a "trance or dream" effect startlingly close to Stevenson's dramatic situation in Dr. Jekyll and Mr. Hyde—but with full recall. This appears to be the heart of his confession.

> *To wake up in the morning and realize what I had done, with a clear mind and all my essential moral and ethical feelings intact at that moment.* [I was] *absolutely horrified that I was capable of doing something like that . . . basically, I was a normal person. . . . I was okay. The basic humanity and basic spirit that God gave me was intact, but it unfortunately became overwhelmed at times. And I think people need to recognize that those of us who have been so much influenced by violence in the media—in particular pornographic violence—are not some kinds of inherent monsters. We are your sons, and we are your husbands. . . . There is no protection against the kinds of influences that there are loose in a society that tolerates.*

Bundy's remarks echo in muted tones Charles Manson's snarls addressed after his conviction to Mr. and Mrs. America. "I am what you have made of me and the mad dog devil killer fiend leper is a reflection of your society." But Bundy was condemning not all society, only a specific segment of it, sexually violent pornography, as it affects a few vulnerable individuals.

In the latter part of the interview, Bundy said he was no social scientist but that he had known a lot of men "motivated to commit violence, just like me. And without exception every one of them was deeply involved with pornography." He cited an FBI report on serial killers. Dobson moved to the crucial questions. "Are you thinking about all those victims? . . . Is there remorse there?" "Absolutely," answered Bundy, and he insisted on returning to the subject of sexualized violence in the media. He said he hoped that those he had harmed would believe his warnings about the effects of those programs. "There are kids sitting out there switching the TV dial around and come upon these movies late at night." Then came Bundy's coda on the punishment he faced.

> *I deserve, certainly, the most extreme punishment society has. . . . What I hope will come out of our discussion is* [that] *I think society deserves to be protected from itself because as we've been talking there are forces at loose in this country—particularly again, this kind of violent pornography. . . .*

We have traveled a long way from the singular, unforeseen, unplanned homicides of Billy Budd and Meursault. But we remain within the territory contained in the phrase "the mystery of iniquity." This was a pathological liar and accomplished con man speaking. We would do well to discount his death-row turn to religion and to watch such tricks as "I was okay." Essentially, he had three things to get across: a plea, a diagnosis, and a warning. Even though it was the only newsworthy item, Bundy's claims of contrition and the higher plea bargaining they represent carry no weight. We have to say: Too little, too late.

His psychological self-diagnosis in the Dobson interview conforms loosely to most early attempts to analyze his case, including his own attempts, using such terms as *compartmentalization*, *alpha and beta personalities*, *the entity*, *the great white shark* for his evil other ego, *personality disorder*, and many more. In the background of these attempts to describe and classify lurks the "to understand is to forgive" syndrome. But Bundy also stoutly maintained—along with almost everyone else involved—that no one could understand him. The mystery of so great depravity lies beyond our capacity to imagine or comprehend.

The warning against violent pornography emerges from the interview as a more straightforward statement than either the plea bargaining or the self-diagnosis. These remarks are consistent with many earlier statements. And in the crucial passage, Bundy changed the emphasis: not violent pornography but pornographic *violence.* Violence becomes sexually arousing. After one hundred pages in *The Only Living Witness* that fill in the background of the case, Michaud and Aynesworth begin to report directly on their "conversations with a killer."

> *Ted began his story with a preamble of operatic sweep and dimensions. . . . His first substantive remarks were on the roles of sex and violence in the development of a psychopath. . . . "this interest, for some unknown reason, becomes geared towards matters of a sexual nature that involve violence. I cannot emphasize enough the gradual development of this."*
>
> (104)

Two vivid pages follow about how violent pornography tends to lead to "the use, the abuse, the possession of women as objects." Subsequent investigations revealed that Bundy probably was privy before the age of four to his maternal grandfather's large collection of pornography kept in a greenhouse. In the April 23, 1978, conversation with Aynesworth, Bundy talked in the agreed-upon third-person style about "the profile we've created" of the serial killer for whom the sex act has changed into a ritualized, impersonal symbol of possession.

> *Uh, with respect to the idea of possession, I think that with this kind of person, control and mastery is what we see here. . . . In other words, I think we could read about the Marquis de Sade and other people, who take their victims in one form or another out of a desire to possess and would torture, humiliate, and terrorize them elab-*

orately—something that would give them a more powerful impression [that] *they were in control.*

(TED BUNDY: CONVERSATIONS WITH A KILLER, 125)

Bundy, like Brady in the Moors murders, had read Sade and about Sade and was speaking here in seemingly objective terms about Sade's effect on him, always the true subject of the conversations disguised behind the third person.

In the Dobson interview, Bundy failed to make an important observation. He maintained that violent pornography criminally influences only a small minority of people, of whom he was one. And he maintained that only a secluded, essentially unknowable segment of himself committed the crimes, while the rest of him remained normal. He did not examine whether it was above all the sick, warped, and depraved part of himself that responded to and was influenced by sexually violent materials. The pat phrase "split personality" does not by itself explain anything. But if we decided to lend some credence to Bundy's last message to a society he had both belonged to and horribly violated, we should perceive the implied suggestion that most normal people can resist corruption but that some temperaments remain profoundly prone to it. We cannot identify that minority by outward appearances, probably not even by professional testing and interview, as Bundy's case demonstrates, until it is much too late.

I find I cannot dismiss Bundy's final interview as no more than public posturing. Like many others, he affirms that violent pornography carries danger for some people—juveniles and a few personality types—and, through them, for all of us. Such statements, added to the evidence we have about Brady's reading materials in the Moors murders case, make a clear contribution to the pornography debate. Where does the burden of proof lie? That is the question we may have to return to. Before restricting pornography in any way as an exception to the protections of free speech, must we demand proof of harmful effects of a criminal nature caused directly by it?* Or, conversely, before allowing unrestricted circu-

---

*The attorney Frederick Schauer points out that we accept so many regulations on speech—such as those on advertising (Federal Trade Commission), on com-

lation and sale of all forms of violent and explicit pornography in the many media of communication now available, should we demand proof that it does *not* have any harmful effects of a criminal nature? How many actual cases and what risks of cruelty and violence, particularly to children and women, should we accept in the name of the principle of free speech? Any satisfactory measure of the effects of images, narratives, and ideas on our behavior probably lies beyond our capacities, another domain of forbidden knowledge, like the mystery of iniquity.

We have before us now two claims. The Marquis de Sade deserves rehabilitation as a great writer and moral thinker. The Marquis de Sade belongs among the greatest immoralists capable of stimulating homicidal madness in warped minds. The time has come to look him squarely in the eye.

5. A Closer Walk with Sade

We exaggerate the sexual appetite in ourselves to take the place of the love we inadequately feel.

—Graham Greene, "Note on Turgenev"

To the reader: In this section, I shall quote and discuss passages that many people will consider offensive and obscene in the extreme. Most writings on Sade and even some anthologies avoid such explicitness and limit their quotations to philosophical discussions of crime, passion, nature, freedom, and the like. To bowdlerize Sade in this fashion distorts him beyond recognition. The actions described in his works directly complement the ideas and probably surpass them in psychological impact. I believe the reader should feel Sade's full effect, however briefly, in order to have a basis on which to consider the issues raised in this chapter.

How seriously should we take Sade? Won't the current intellectual fashion regarding his work blow over like that of, say, Lavater's phrenology in the past century? Does it make any difference that

mercial behavior (the securities laws), and on perjury—that free speech could be seen as an exception to the wider principle that communication may be restricted.

both Ian Brady and Ted Bundy had read Sade? The interpretations and opinions traced in my earlier section on his rehabilitation suggest how strong his presence has grown at the end of the twentieth century, particularly among intellectuals.

At the beginning of this chapter, I called Sade a "test case." I can now restate that proposition as a general question and as a particular question. Shall we receive among our literary classics the works of an author who desecrates and inverts every principle of human justice and decency developed over four thousand years of civilized life? Has the twentieth century made, in respect to the Marquis de Sade, one of history's most egregious errors of cultural judgment by placing his works among our literary masterpieces? Many readers will probably bridle at such peremptory challenges and brush by impatiently, thinking, He cannot be that bad. There's no need to be so judgmental. Meanwhile, a much smaller group of scholar-readers is deciding those questions for the rest of us. We need to examine the evidence.

One preliminary observation needs to be made. Large segments of Sade's writings give an impression of late-eighteenth-century Romanticism spun out of contemporary travel accounts. They seem to contradict the deliberate offensiveness of his better-known works. Except for a few descriptions of savagery in Africa, *Aline and Valcourt* would bore more readers than it would shock. It even contains a description of Sade's utopia, Tamoé, hidden on an island off New Zealand. King Zamé, who has visited and gained knowledge from Europe, runs Tamoé as a socialist despotism based on equality of living conditions for all, state care of children from birth, temperance and benevolence, and light punishment for what little vice and crime occur. Surprisingly parallel to Thomas More's *Utopia,* these pages have an earnest tone, with none of the satirical twists that give a pungent flavor to *Gulliver's Travels.* The libertine segment of Sade's work, however, offers many examples of a very different kind of society: In an underground bunker protected by moats and gates, a man or small group of men create a safe and luxurious environment for total mayhem, sexual and homicidal, inflicted by a few masters on dehumanized victims.

Two samples brought up from deep inside Sade's universe will have to represent him here. They can only suggest the extensive surrounding terrain. I have deliberately chosen them from extreme situations because his major works carry the reader fatefully toward such confrontations with inhumanity. We already have a word to designate the place we are now entering: *taboo.* Frazer defined primitive taboo as "holiness and pollution not yet differentiated." The most convincing version of that tension comes in Eve's double reaction to the Tree of the Knowledge of Good and Evil as praised by the serpent: fear and fascination. They work all her woes, and Adam's after hers. Sade's writings exploit a new form of what I shall call "civilized taboo." After four millenia of religion and philosophy and statecraft have gradually differentiated between holiness and pollution, Sade sets out to confound them again. By manipulating fear and fascination, he tries to confer holiness on our most deeply polluted impulses, and vice versa. Anyone who does not register a sense of taboo in reading Sade lacks some element of humanity.

Written in the form of a dialogue, *Philosophy in the Boudoir** relates the systematic initiation of fifteen-year-old Eugénie by her twenty-six-year-old woman friend, Madame de Saint-Ange; by Madame de Saint-Ange's younger brother and partner in longstanding incest, the Chevalier; and by Dolmancé, thirty-six, "a sodomist on principle" who scorns ordinary lovemaking. By preachments, example, and formal lessons (including inspection and explanation of Dolmancé's "member . . . the principal agent of love's pleasure" and of Eugénie's clitoris, vagina, and anus), they introduce Eugénie into full participation in their orgies. The timeouts are filled primarily by Dolmancé's philosophical disquisitions to justify sodomy, cruelty, and murder and to condemn propagation of the species as contrary to nature. Children are to be eliminated like nails and excrement. After much mutual masturbation and buggery (the women use dildos), the Chevalier deflowers Eugénie in an elaborately choreographed tableau. She is immediately refucked by a valet with a mammoth cock, who is called in for the occasion.

The following intermission is occupied by the reading aloud of a revolutionary pamphlet, "Yet Another Effort, Frenchmen, If You Would Become Republicans." Its forty pages describe a totally libertine state of society permitting every excess and crime, including murder but excluding capital punishment and war. Dolmancé sums

*Not "Bedroom": There are no beds in Sade, only couches and sofas.

up by dismissing pity and benevolence as the most despicable of human qualities. The Chevalier, not yet totally depraved, responds with a spunky speech defending purity of heart, sensibility, and the sacred voice of Nature calling us to goodness. Dolmancé scornfully affirms that Nature is everywhere cruel and ungrateful, and that "heart" refers only to "the mind's frailties." "One unique flame" of selfish passion brings "the inestimable joys that come of bursting socially imposed restraints and of the violation of every law." Eugénie declares that Dolmancé has won the argument and takes the initiative in her initiation by asking Dolmancé about "the cruelty you recommend with such warmth." Paraphrase will no longer suffice.

EUGENIE: *Please tell me, how do you view the object that serves your pleasure.*

DOLMANCÉ: *As absolutely null. . . .*

EUGENIE: *Why, it is even preferable to have the object experience pain, is it not?*

DOLMANCÉ: *To be sure. . . . The repercussion within us* [in response to their pain] . . . *more promptly launches the animal spirits in the direction necessary to voluptuousness. Explore the seraglios of Africa, those of Asia, those others of southern Europe, and discover whether the masters of these celebrated harems are much concerned, when their pricks are in the air, about giving pleasure to the individuals they use; they give orders, and they are obeyed; they enjoy and no one dares make them answer; they are satisfied, and the others retire. Amongst them are those who would punish as lack of respect the audacity of partaking of their pleasure. The king of Acahem pitilessly commands to be decapitated the woman who, in his presence, has dared forget herself to the point of sharing his pleasure, and not infrequently the king performs the beheading himself. This despot, one of Asia's most interesting, is exclusively guarded by women . . .*

There is not a living man who does not wish to play the despot when he is stiff: it seems to him his joy is less when others appear to have as much as he; by an impulse of pride, very natural at this juncture, he would like to be the only one in the world capable of experiencing what he feels: the idea of seeing another enjoy as he enjoys reduces him to a kind of equality with that other, which impairs the unspeakable charm despotism *causes him to feel. . . .*

The debility to which Nature condemned woman incontestably proves that her design is for man, who then more than ever enjoys his strength, to exercise it in all the violent forms that suit him best, by means of tortures, if he be so inclined, or worse. Would pleasure's climax be a kind of fury were it not the intention of this mother of humankind that behavior during copulation be the same as behavior in anger? What well-made man . . . does not desire . . . to molest his partner during his enjoyment of her? I know perfectly well that whole armies of idiots, who are never conscious of their sensations, will have much trouble understanding the systems I am establishing; but what do I care for these fools? . . . soft-headed woman-worshippers. . . .

Goddam! I've got an erection! . . . Get Augustin [the valet] *to come back here, if you please.* (They ring; he reappears.) *"Tis amazing how this fine lad's superb ass does preoccupy my mind while I talk! All my ideas seem involuntarily to relate themselves to it. . . . Show my eyes that masterpiece, Augustin . . . let me kiss it and caress it, oh! for a quarter of an hour. Hither, my love, come, that I may, in your lovely ass, render myself worthy of the flames with which Sodom sets me aglow. Ah, he has the most beautiful buttocks . . . the whitest! I'd like to have Eugénie on her knees; she will suck his prick while I advance; in this manner, she will expose her ass to the Chevalier, who'll plunge into it, and Madame de Saint-Ange astride Augustin's back, will present her buttocks to me: I'll kiss them. Armed with the cat-o'-nine-tails, she might surely, it should seem to me, by bending a little, be able to flog the Chevalier who, thanks to this stimulating ritual, might resolve not to spare our student.* (The position is arranged.)

(*JUSTINE . . . AND OTHER WRITINGS*, TR. RICHARD SEAVER AND AUSTRYN WAINHOUSE, 343–46)

Dolmancé ends up deliriously sucking the Chevalier's semen out of Eugénie's dripping ass.

In the final scene, Eugénie's mother arrives to rescue her. The four libertines mercilessly rape, sodomize, and torture the woman and then have a syphilitic valet infect both her orifices. Eugénie finally brings herself to orgasm by sewing up her mother's vagina and ass amid spurts of blood. The pupil has outstripped her instructor. She has almost become a man.

I know two responsible readers who consider *Philosophy in the Boudoir* effective in a backhanded way. One of them finds the work so extreme and horrible as to inspire revulsion from such behavior and from the principles advanced to justify it. The other reader finds the situations so exaggerated and grotesque as to provoke derisive laughter. For both, the dialogue works entirely by recoil, like a hellfire sermon or an old army film on venereal disease. (The latter reader dismisses all Sade's other writings as overly long and unreadable.) Evidently, the book affects readers in different ways. How many, if drawn in, will seek cruelty and violence to augment sexual pleasure?

Boudoir is one of the few of Sade's works in which one can glimpse moments of incipient situation comedy. But they are more lapses in style than deliberate effects. Irony never diverts or relieves Dolmancé's sermons on systematic criminal egoism. Incredible as it may seem, he and Sade mean what they say about immorality, torture, tyranny, and wholesale murder. This is no flippant game out of Oscar Wilde, no dirty joke from the barracks. Barthes' no-fault interpretation of Sade cannot remove the appalling moral burden from this short dialogue. If the Moors murderers and Ted Bundy furnish any guide, the taboo effect—fascination and revulsion—of such a work can be extremely powerful on some people, particularly among the young, the unbalanced, the criminally inclined. Such minds cannot purge so searing a message. It works within them like a personal challenge, daring them to act accordingly. And they may come to believe, as Brady virtually stated under cross-examination, that arguments like Sade's legitimize torture and murder.

The succinctness of *Philosophy in the Boudoir* (two hundred pages) is not matched by any of Sade's other major works. *Justine*, whose several versions run to thousands of pages, exploits the titillating narrative device of a young woman who remains morally chaste and innocent through the most horrible violations of her body in the hands of a succession of depraved males. Justine offers us a two-dimensional version of the already implausible Clarissa.

Her sister, Juliette, in the immense picaresque novel that bears her name, takes the opposite course. Originally initiated into lesbian debauchery in a convent, she practices libertinism and prostitution to satisfy her passions, dominate others, and amass a series of fortunes. Juliette almost but not quite succeeds in exceeding in depravity the males whose favorite she becomes. After seven years of travels and debaucheries, she returns to Paris and links up in a mammoth orgy with her first master, Noirceuil, who originally ruined and killed both her parents. For him, "crime is the soul of lubricity," and he justifies his outrages by a philosophy that simply classifies them as a part of nature's design. Born noble, rich, and powerful, Noirceuil has recently murdered his best friend and protector in order to replace him as prime minister. The conspiracy fails. Noirceuil's eleven-inch prick is his only god. "Let it be yours too, Juliette, this despotic cock. Do full honors to this superb deity. I want to expose it to the homage of the whole world." At the end, Juliette is also reunited with her seven-year-old daughter, who is included in the final orgy along with Noirceuil's two teenage sons, who have been deliberately brought up as complete savages.

After a bizarre double wedding ceremony in drag among members of the same sex, Noirceuil and Juliette barricade themselves in his château for the great bacchanale with her daughter, his sons, two torturer-excutioners, and half a dozen victims of both sexes. Their pleasures are fed by the most unspeakable humiliations and outrages to the participants. The sons are forced to bugger the father, who imitates the shrieking behavior of a young virgin. Whippings begin, blood flows, breasts are ripped off, limbs are broken and dislocated, and eyes are torn out while Noirceuil sodomizes the victims and has Juliette fucked front and back by obedient flunkies. Sade writes very graphically. Brought to extreme arousal by the excruciating torture to death of two female victims, Noirceuil buggers one of his sons while literally eating the boy's heart, which has been torn out of his body by Juliette. The performance continues as narrated by Juliette herself in the historical present.

> *His eyes blazing, Noirceuil now falls on my daughter; he has a monstrous hard on. He seizes her, over-powers her, and encunts her upside down. What do you want to do with her, Juliette? Could you be such an imbecile as to have any feelings, any concerns for this*

disgusting product of the sacred balls of your abominable husband? Sell me the slut, Juliette. I'll pay for her. I want to buy her. Let's besmirch ourselves together—you by the sin of selling her to me, I by the even more stimulating sin of paying for her in order to kill her. Oh, yes, yes Juliette. Let's murder your daughter!" And pulling out his prick to show it to me: "Examine how far this execrable idea inflames all my sense. Get yourself fucked now, Juliette, and don't answer me until you have two pricks inside you."

Crime has nothing fearsome about it when you're fucking. And it's always in the midst of waves of come that one must cherish its charms. I'm being fucked. Noirceuil asks me a second time what I want to do with my daughter.

"Oh Monsieur," I cry out while discharging, "you win, your perfidious power smothers every sentiment in me except crime and infamy. Do what you want with Marianne, blasted villain. I turn her over to you."

He had no sooner heard these words then he pulls out of her, grasps the poor child, and throws her naked into the middle of the flames. I help him with a poker to arrest her natural compulsive responses to save herself by coming back to us. Others are diddling us and buggering us. Marianne is roasting. She is consumed. Noirceuil discharges. I do the same. We go to spend the rest of the night in one another's arms, congratulating ourselves on the scene whose episodes and circumstances complete a crime that still does not satisfy us.

"Well now," Noirceuil says to me, "is there anything in the world worth the pleasure of crime?"

"Oh, my friend, I know of none."

"Let us live eternally that way. May nothing in all nature ever carry us back to different principles. Happiness depends on the strength of one's principles."

(JULIETTE, ŒUVRES COMPLÈTES, Tome 9, 576–77)

For a week in the security of Noirceuil's estates, they practice every bestiality and systematically deprave schoolchildren by means of buggery and mutual torture before poisoning them en masse. Finally, Juliette's imagination surpasses Noirceuil's. While masturbating him, she persuades him to poison fifteen hundred people, whose deaths will be attributed to an epidemic. At the end of Juliette's account of this most recent episode in her story, her prudish sister is sent out into a violent thunderstorm and the natural force of lightning kills her by destroying her cunt and leaving her anus intact. Aroused by Juliette, the four libertines in the group, including Noirceuil, gang-bugger the corpse while Juliette diddles herself.* Now, on the last page of the novel, a courier arrives from the King at Versailles to announce the appointment of Noirceuil as prime minister, in which position he is to take over absolute powers of government. Noirceuil triumphantly declares that his appointment represents the reward of vice and the punishment of virtue—though "we might hesitate to say so if we were writing a novel." Do those words represent a self-conscious literary jest that neutralizes everything and undoes all the damage? No, for Juliette herself has the last word about the commanding moral tone of what we are reading. "Why be afraid to publish the secrets that truth itself rips out of nature. . . . Philosophy must say everything."

The passages just quoted from *Philosophy in the Boudoir* and *Juliette* allow me to make a number of general remarks about Sade's most widely read works. His situations arise from the existence of a rigid caste system, that of the Ancien Régime in France. The male characters are wealthy, powerful, usually of noble birth, owning vast estates on which to behave as tyrants and to carry out their debaucheries in safety. Sade's one revolutionary pamphlet and his island utopia of Tamoé show some concern for ordinary citizens. But in his fiction, the total license to take one's pleasure anywhere at will extends only to a tiny minority of rich and powerful aristocrats. Everyone else must be a victim. Verneuil gives the doctrine full expression in *La Nouvelle Justine, VII.*

> *It is impossible that laws should apply equally to all men. This moral medicine is no different from physical medicine: wouldn't you laugh at a quack who, having only one remedy for all customers, would purge a stevedore the same way as a flighty spinster? Of course! Laws are made only for the common people: being both weaker and more numerous, they need restraints that have nothing*

*This version reverses the ending of *Justine* described on page 233.

to do with the powerful man and that do not concern him. In any government the essential thing is that the people never invade the authority of the powerful.

(CHÂTELET, 121)

Wives must accept the role of slaves. Sade's heroines are universally beautiful, young, well shaped, indestructible even after prolonged abuse and disfigurement, and generally compliant. In *Philosophy in the Boudoir* and *Juliette*, however, Sade begins to modify women's position in his plots and in society. Through wiliness and brashness, a few rare female victims strive to become superior to their male masters, to become victors. One of the lesbian nuns in the first part of *Juliette* has a three-inch clitoris "destined to outrage nature" by buggery. "She's a man." Several female characters later in the novel "have an erection" *(bander)* and "ejaculate sperm." Women and men can never be equals unless they have formed a single sex devoted to sodomy. In all this, Sade reinforces the system of rigid social castes and seeks insofar as physical difference will permit a certain dissolution and homogenization of the sexual. Sade's promotion of women's roles and power in *Juliette* further excludes sentiments of tenderness and intimacy between people of any sex.

The philosophical disquisitions that occupy about half the space in Sade's works of fiction develop a defense of crime and vice as a necessary part of nature in general and man in particular. No violent, homicidal behavior should be condemned as wicked or criminal. Time after time, the justifications of rape and murder as natural and inevitable sound like Pangloss' justifications of wars and earthquakes as belonging to the inscrutable order of Providence. In theory, Sade has changed every sign from minus to plus. But the words themselves never change. He maintains the vocabulary of condemnation and moral outrage. The words *horrible*, *monstrous*, *villainous*, *infamous*, and the like never cease to spice up his sentences. The transvaluation of all values does not find a new language; rather, it reinforces the sharp edge of conventional terms in order to underline the scandal of what is described. And for ordinary people, no change has occurred except to grovel under a new justification of their exploitation by the powerful. Saint-Fond explains, "Everything called licentious crime, such as murder in the course of debauchery, incest, rape, sodomy, and adultery, will never be punished except in the slave castes." The laws do not govern the highest caste of elite libertines; laws only reinforce their superiority.

The principal metaphor that expresses this moral and social system is fire. Sade relied on it all his life. In *Philosophy in the Boudoir*, Dolmancé summons Eugénie "to drown in floods of fuck the heavenly fire that blazes in us." But the fire never dies. A few pages later when she has some scruples about infanticide, Dolmancé replies that "the torch of philosophy has dissipated all those impostures." By the time we reach the passage from *Juliette* quoted on pages 274–75, real flames are burning right there onstage to symbolize the all-consuming lusts of Noirceuil and Juliette and to destroy both her daughter and any lingering remnants of maternal attachment to her child. Fire speaks a single message: My lust will destroy all. The intensity of my orgasm (a word not available to Sade) reduces the rest of the world to being my victim.

The two passages quoted only begin to suggest the degree to which Sade relies on cliché and stereotype to describe his characters and set his scenes. All women are "beautiful as Venus" or "lovely as the day"; superlatives tumble out in series to establish the terrible power of the men, particularly their enormous cocks. Size and number will overcome boredom. Sade wants a cartoon world of two-dimensional figures unfettered by reflection or remorse. Their abundant conversation justifies their conduct. When he wishes to imply further reaches of meaning, Sade usually falls back on one factotum adjective: *interesting.* Saint-Fond's enslaved wife receives the designation "that interesting creature." And that's the end of it.

Inevitably in such a comic-strip universe, moments of comedy burst out. The innocent Eugénie keeps running ahead of her corrupt tutors. Candidates for the terrible ordeal planned at the opening of *The 120 Days* are screened like astronauts and given names worthy of Rabelais: Rip-Ass, Peck-Stiff. To this universe awash in fire, sperm, and blood, why isn't our response an immense guffaw at the exaggeration of everything, the sheer preposterousness of anyone ever even beginning to carry out such exploits? The English poet Algernon Swinburne apparently reacted this way on first reading Sade in 1862. He could have "died laughing" at Sade's naïveté in taking "bulk and number for greatness." Later, Swinburne drew heavily on Sade's emphasis on pain. Apollinaire was also half-inclined to this Rabelaisian or Ubuesque response. But, considering the stereotypes Sade employs, why is the dismissal by laughter so rare?

The explanation lies, I believe, in Sade's reliance on what could be called "the *Bolero* device." With small variations in instrumentation and key, the same motif is repeated over and over and over with increasing intensity until it has tattooed itself on the mind. Philosophical discourse on evil and selfishness alternates with enactment of these ideas. Both are intended to be prurient. It is as if Sade had set out to disprove what Proust wrote a century later: "Nothing is more limited than pleasure and vice." The narrative goes on and on, with small increases in depravity. The one stylistic effect Sade has mastered is the crescendo. He knows how to turn the volume up slowly. Instead of laughing at the *Bolero* device, we tend to find other responses. We are aroused; Sade does not hide the fact that this is his preference and his purpose. Or we are simultaneously fascinated and repelled, a standard reaction that defines taboo. Or the writing deflects us into uneasy reflection about the significance of such grotesque stories. Or we lapse into boredom, broken at intervals by new horrors.

Sade's insistent *Bolero* device and his preferred situation of initiation and instruction combine into a single effect. His writing is didactic. He seeks to convince and convert—himself first of all, one feels, then all the rest of us. In this respect, Barthes was right to link Sade with Loyola and Fourier, two great proselytizers with a doctrine to preach. Sade was too much of a fanatic in his writings to provoke laughter. *Philosophy in the Boudoir* presents a catechism class in evil and debauchery. *Juliette* enacts the triumph of the new doctrine by transforming victim into master, woman into man. One of Juliette's masterful lesbian friends, Madame de Clairwil, seeks "a crime whose effect will continue even if I stop . . . while I'm asleep." Juliette answers her that advice, writing, and action are three ways of sowing a powerful "contagion" that will last forever. We quickly understand that publishing books like Sade's own works combines all three activities.

To what did Sade hope to convert us? As I have pointed out

earlier, we cannot believe his early protests that he paints vice in order to uphold virtue. The works themselves belie him as well as later, more forthright statements of intent. No, he preaches evil, selfishness, and destruction and sets them before us as vividly as he can. A contemporary scholar of French literature, Jane Gallop, spends twenty pages arguing that the central Sadean doctrine is the primacy of the act of sodomy. He and she find buggery more radical, more violent, "more zestily criminal" than vaginal intercourse. "Anal intercourse is the keystone of a system enabling the individual pervert, locked into his singularity, to engage in a generalized exchange." Behind this dubious psychoanalysis, Gallop is perfectly correct in insisting on the all-pervasiveness of sodomy in Sade's writings. Camille Paglia says it more directly: "Sodomy is imagined as ritual entrance to the underworld, symbolized by a man's bowels."

But surely Sade's evangelism seeks to convert us to a larger faith than buggery. To pursue this point, we must face a major question of consistency and contradiction in Sade that draws us to the center of his ethos and to the fascination he seems to exert in our time, more on intellectuals than on mere seekers of prurient interest. Another scene from *Juliette* clarifies the terms of the dilemma. At a secure country estate, Juliette imprisons a worthy family of three, whom she has instructions to hold for execution. Soon she welcomes Decour, the well-built official executioner who will do the job. At dinner, Decour explains that murder represents simply a necessary step in nature's process of destruction and regeneration.* Yet he admits that he gets a hard-on when executing someone. The discussion leads to extended sex and buggery interrupted by frantic whippings, until finally Juliette can produce the needed formula for the occasion. "When one becomes accustomed to scorn the laws of nature on one point, one cannot find any pleasure unless one transgresses all of them one after the other."

Something is amiss. Murder was just presented as a necessary part of nature's process. Now murder belongs to generalized transgression. That contradiction falls within a larger one. On the one hand, if we follow the essential nihilism of nature, there are no

---

*We are not told how regeneration can occur if propagation and offspring are contrary to nature (see p. 270).

laws or moral principles to restrain us from total debauchery and destruction. On the other, life and particularly our pleasure in it lose their intensity in such a void, deprived of those defining laws. Beyond a certain point, the inveterate libertine cannot attain orgasm without a sense of transgressing the laws and constraints that he rejects. The door must be both open and closed. In the first chapter of *Juliette*, the lesbian mother superior of the convent flaunts to Juliette her "epidemic licentiousness," whose destructive force belongs to nature's purpose. And she explains that the faculty we call *conscience* is the work of useless prejudices. One can construct a contrary conscience that urges one to commit every excess without remorse. But somehow everything has stayed in place. "Remove the punishment, change public opinion, destroy the law, renovate the thinking subject, and crime will still be there, though the individual will feel no remorse."

Why haven't the signs simply changed now? Why hasn't crime become virtue, if it represents nature's way, and vice versa? At least "crime" should be redesignated by neutral terms unassociated with moral judgment. But Sade simply cannot give up the limits he goes to such lengths to desecrate by actions and by philosophical demonstration. In order to attain the needed sense of crime and excess, all Sade's libertines remain parasitic on the constraints they deride. Without limits, our actions could have no naughtiness. The innocent orphan Justine is instructed by one of her first exploiters that "God is what interests us least in the world" and in the same sentence that "our passions have no charm except when they transgress His intentions."* We call this wanting to have it both ways. Sin, even when assigned a plus sign as part of natural process, must remain sin with a minus sign in order to assure us our moral kicks. Sade the missionary of transgression tolerates laws and limits in order to be able to trample them underfoot. We expect such willful naughtiness in children and in some disturbed mental cases, not in adults. A facetious version of it crystalizes in St. Augustine's quip that, yes, he was resolved to confess and reform his sinful ways—

*A variation on this attitude led Michel Foucault at the end of his life to claim and perhaps to believe that his S&M debaucheries represented a form of ascesis, a heroic philosophic experiment. James Miller's book on Foucault describes this distressing case and, I believe, misinterprets it.

but not quite yet. Sartre unmasked Baudelaire's dependency on this piquant notion of sin, and in the process gave currency to the term *bad faith* as hypocrisy toward oneself.

Two clearheaded and well-informed books complete my narrow discussion of transgression in the wake of Sade. In *The Romantic Agony* (1931), Mario Praz surveys authors of the Decadent school in England and Europe who exult in evil. Susan Brownmiller's *Against Our Will: Men, Women, and Rape* (1975) confronts us with the tenacious tradition of the male outlaw, to whom every excess is permitted in the name of conquest and defiance. Between them, they provide a whole curriculum of transgression, of evil affirmed as the prerogative of the strong, from Richardson and Laclos through Huysmans to the Kubrick movie of Burgess' *A Clockwork Orange.* Unfortunately, Praz and Brownmiller virtually ignore the author who has become the principal standard-bearer of Sadean nihilism. In Nietzsche, the ethic of transgression has been stripped of scenes of explicit sexual torture and destruction and raised to an alluring intellectual Valhalla of lyric philosophy. We have no evidence that Nietzsche ever read the divine marquis. Nevertheless, the philosopher of the Overman offers a modified product with enhanced appeal for some: Sade without orgasm.

Sade's writings confront us with the extreme attempt in Western culture to strip away the constraints of civilization in order to return to barbarism. In all his major writings, Sade envisages a complete rejection of Hebrew law and prophecy, of Greek philosophy and tragic vision, of Christian charity and service, and of all principles of equal justice and democracy. He seeks to revive the talion law of an eye for an eye and might makes right. Perhaps there is something "great" in the sheer atrocity of Sade's work, some monumental aberration and object lesson that we should hold in awe. But it will seem less admirable if we read him whole and keep his nihilistic ideas about egoism and power closely tied to his lurid scenes dripping with blood and feces. He knew "the importance of these pictures to the soul's development" and how "to lay hands fearlessly on the human heart and portray its gigantic divagations" *(Justine)*. Sade was always the teacher and evangelist.

6. "MUST WE BURN SADE?"

Simone de Beauvoir's question provokes us by alluding to the extreme measures of the Inquisition. But I do not believe the question should be dismissed or left unanswered, even though no form of book burning or censorship could eliminate Sade's writings—let alone his myth—from libraries and private collections, from the historical record, and from collective memory. His profusely illustrated moral nihilism has entered our cultural bloodstream at the highest intellectual and at the lowest criminal levels. Sales figures provided by his American publishers confirm these statements. Between 1965 and 1990 the 750-page edition of *Philosophy in the Boudoir* and *Justine* sold 350,000 copies and now averages sales of about 4,000 a year. The companion volume, *The 120 Days of Sodom*, sells somewhat fewer copies. These are substantial numbers. They also represent substantial profits.

No, we must not burn Sade. I come to that conclusion not because it would be impossible to do so but because we should not deliberately destroy any human life or accomplishment, even the most excessive and monstrous. Medical laboratories preserve the most virulent strains of fatal diseases for educational and research purposes. But let us not stop here. Beauvoir's question does not address the fundamental issue.

The right question is more timely and more defiant: Should we rehabilitate Sade? Should we rank him as a major thinker and writer to read along with Machiavelli and Rousseau? George Eliot and Dostoyevsky? Should we follow the Harvard *History of French Literature* in celebrating his work as "the triumph of desire over objective reality"? In my pages on "Rehabilitating a Prophet," I rebut the four basic claims made to establish Sade's stature: He had a powerful imagination; his works have importance as scientific documents; he was a great revolutionary; and he articulated an original and significant moral philosophy. Each claim has limited application and is seriously flawed from a literary and philosophical standpoint.

One more circumstance, both biographical and literary, further weakens all four claims. For all his advocacy of sodomy as the supreme act of transgressive pleasure, Sade does not conceal that his truest (and for long years enforced) allegiance was to auto-eroticism, physical and mental. Increasingly with age, Sade kept count in his journals of his own sexual activity, including masturbation. He also documented the behavior of his fictional characters. At the height of her material success as a criminal, Juliette goes periodically to masturbate while contemplating her immense fortune in gold, which symbolizes "crime at my disposition." Early in Part IV of the same novel, Juliette carefully instructs the Countess de Donis about her "secret method of inflaming the imagination to break all barriers of conduct. After at least two weeks of abstinence one should masturbate very deliberately and let one's mind imagine in detail the most horrible satisfactions and exciting outrages." One should take note of these imaginings for later application and at intervals repeat the operation with increasing intensity. Juliette guarantees the results.

Having considered Sade's case with great care, Camus drew the unavoidable conclusion. "Prometheus ends up as Onan." Sade was more a prisoner of his masturbatory phantasms than of the Bastille or Charenton.

We have come to one of the largest of all questions: How shall we talk about moral questions? If we accept the authority of a revealed faith or of an established tradition, we can appeal to commandments and principles handed down and accept their guidance in particular instances. If we do not accept such authority, or at least not in the area under consideration, we can still work as Aristotle and Cicero did with practical argument *(phronesis)* based on the comparison of exemplary cases. This approach relates closely to common law, to common sense, and to the clinical method in medicine, for *phronesis* implies that moral knowledge is more particular than theoretical. What it requires is acquaintance with relevant history. The virtue of the case method, properly employed, is that it avoids the extremes of dogmatism in theory and of relativism in mere description.*

---

*Within the extensive library of contemporary writings on moral philosophy (including such serious thinkers as Alasdair MacIntyre, Iris Murdoch, John Rawls, Paul Ricoeur, Charles Taylor, and Bernard Williams), the most useful and trenchant discussion for my purposes is a historical study. In *The Abuse of Casuistry: A History of Moral Reasoning*, Albert R. Jonsen and Stephen Toulmin describe the contentious development of the Catholic church's dealing with moral behavior and its unjustified discrediting by Pascal's Jansenist attacks. Insofar as casuistry remained an

What, then, are the cases we should invoke in order to gain a wider perspective on the life and writings of Sade? Where else do we find extensive writings that mix stories and instruction in an attempt to modify the sexual and moral behavior of a culture? Barthes' association of Sade with Loyola and Fourier is too limited and moves in the wrong direction. Sade's own references to influential figures ranged over a much vaster field of history and geography. I shall cite a handful of what I believe to be pertinent cases. They take us far from the scandals surrounding Sade, but I shall not be straying from the subject. The divine marquis belongs in a wider context than that of eighteenth-century pornographic novels.*

Under the early Han dynasty in China (206 B.C.–A.D. 24) Taoist doctors and scholars formulated an extensive set of classical sex manuals often referred to as *The Art of the Bedchamber.* As Robert H. van Gulik informs us, the sexual encounter had both cosmic and personally therapeutic consequences if performed with proper controls over female yin and male yang. *Secrets of the Jade Chamber* goes to great lengths to describe how to enhance and to prolong sexual pleasure. These manuals enjoyed renewed popularity in a period of sophisticated excess at the end of the Ming dynasty during the closing years of the seventeenth century. The explicitness of the manuals is veiled by elaborate euphemisms—such as "jade stalk" and "red flower"—and cleared up by the illustrations. Though widely suppressed under the Manchu dynasty, these manuals have had a lasting affect on sexual practices and sexual morality in China.

In comparison with the obscure Taoist manuals based on scholarship and tradition, Ovid's world-famous *Ars Amatoria* reads like a jaunty instruction manual on flirtation and seduction in respectable society. Ovid's narrator connives, takes chances, preens, and mocks himself, all the while dispensing good-natured advice free of sexual obsessions. The brief pages at the end of Books Two and Three

---

honest case method dealing with lived experience, it represented an admirable attempt to reconcile principles and practice. The opening and closing chapters of Jonsen and Toulmin's book speak revealingly of contemporary questions precisely because their historical foundation is solid.

*The most informative and least tendentious history of sexuality I have found is Reay Tannahill's *Sex in History* (1980).

on how to disport oneself in bed counsel genuine attentiveness to one's partner and a commonsense approach to positions, timing, and pillow talk. Ovid's racy verse plus his often jocular images (military and agricultural) create a world that associates love, pleasure, and laughter. Ovid's perennial appeal springs from his art of creating a lyrical bawdy without coarseness.

Another great tradition of sex instruction was compiled in the Hindu *Kama Sutra* some time after the third century A.D. It covers everything from technique in copulation to proper social behavior and the complications of what we would call romantic love. Tantrism added a further layer of magic, religion, and ritual to the cultivated pursuit of sexual pleasure.

The comparable accomplishment of western Europe in this enterprise of associating sexual practice with religious beliefs wears a somewhat different aspect. Out of the Crusades, the Catharist heresy, troubadours' love songs for an unattainable lady, and the *chansons de geste,* the court of Eleanor of Aquitaine developed the institutions we know as courtly love and chivalry. Andreas' *The Art of Courtly Love* tells how the knight should strive to earn his lady's favor and ask no more of her than a word of commendation. *Amor purus* remained to a large extent a myth or ideal, yet one with powerful effects through several centuries of European history.*

If sexuality is partly absorbed by spirituality in courtly love, revolutionary politics and anticlericalism lay claim to sex in eighteenth-century French pornography. The works of Nerciat, Mirabeau, and Restif de la Bretonne and many anonymous titles such as *Thérèse philosophe* (1748) combine philosophical exposition and sex stories to enlighten readers on the new libertinism of mind and body. Books devoted to erotic behavior became a graphic demonstration of revolutionary ideas, and also earned good money. This outburst of pornography served as a vehicle for attacks on religion, the monarchy, and the aristocracy and provided a form of sex education along with a lot of entertainment. In varying degrees, the

---

*Aretino, the sixteenth-century Italian "scourge of princes" often identified as the first pornographer, has a small claim to a place in this series. Much of his work was political satire. His *Ragionamenti,* dialogues between an experienced and an innocent woman, and his *Sonnetti lussuriosi,* composed to accompany woodcuts of the positions for lovemaking, explored a new level of explicitness for the age of the printing press. His indifference to moral concerns leads me to leave him out.

authors had political ends in mind. After the Revolution, pornography was in part replaced by other institutions and entertainments, such as novels and the theater. The rediscovery and republication of these works in the 1990s has created a juicy new scholarly specialty and led to some overevaluation of their historical and literary importance.*

The last set of writings I shall cite, *Thousand and One Nights*, was partially assembled by Western translators beginning in the eighteenth century from a corpus of Near Eastern and Indian tales in the Arabic language. The frame story bears directly on our subject. The wanton behavior of his wife and concubines provokes the Sultan Schariar into punishing them all with strangulation or torture. Thereafter, Schariar's vizir procures for him each night a new "wife" to be strangled the following morning in order to forestall any unfaithfulness. We are not told whether Schariar finds any pleasure in this barbarous practice. Finally, the vizir's own daughter, Scheherazade, asks to be chosen as the Sultan's wife for a night. She puts an end to the scourge by beginning so absorbing a story in the morning that the Sultan asks for her again the next night in order to hear the continuation. Scheherazade makes this narrative feat work for 1,001 nights, after which the Sultan spares her life for love of the three beautiful children she has borne him.

I am proposing that these writings, different as they are from one another, follow an impulse to modify sexual behavior as a means to influence and presumably to improve the social life of a culture. They envision a better state of things through sexual instruction. A comparison of these works with Sade will now reveal one item of crucial importance to this inquiry. Sade wants to change the world as much as they do—probably more. But he alone makes the highest pleasure of body and mind depend on violence and torture directed toward one's partner or partners and on perversions centered on anal intercourse. The Taoist *Art of the Bedchamber*, Ovid's *Ars Amatoria*, the *Kama Sutra* and Tantrism, and courtly love do not anywhere appeal to practices we would call sadistic or masochistic in order to attain sexual fulfillment. Eighteenth-century French pornography stops at nothing, but it does not make cruelty

---

*Robert Darnton and Lynn Hunt provide sound overviews without entirely resisting inflationary pressures.

its primary drive. Insofar as punishment and slaughter are associated with sexuality in *Thousand and One Nights*, that extended narrative reveals how the Sultan is cured of his cruelty and learns a form of family-directed love. Several of these writings recognize nonviolent behavior like fellatio, cunnilingus, acrobatic postures, and anal intercourse as minor deviations to be acknowledged, not glorified. The Marquis de Sade differs from the others—I hesitate to call them his predecessors—precisely in the meaning of the word forged on his name. He is the only sadist. This systematic association of sexual gratification with malevolence, pain, torture, and murder is something new in social history. Until the nineteenth century, no one needed a word for it. Here is the "mutation" Foucault identified and celebrated. From here springs the consecration of a new literary classic we have chosen for our progeny.

There have of course been many individual cases of men and women who in their sexual conduct reveled in vicious cruelty. A few became famous by their excesses. We remember Nero and Gilles de Rais and Countess Erzsebet Bathory and perhaps Lord Castlehaven, a scandalous contemporary of Milton. In our own time, we have more figures than we want to recall, from Jack the Ripper to Jeffrey Dahmer. Most of these are pathological cases. No one has proposed that any of them furnish an example in whose name we might organize society in order to reach higher levels of human fulfillment. The rehabilitation of the Marquis de Sade, however, appears to call for precisely that goal. The passages I have quoted from many critics and scholars treat him as exemplary. Can they be serious? Do they know what they are doing?

The introduction by Michel Delon to the Pléiade edition of Sade in France addresses precisely these questions. By quoting from many of the sources I have already discussed, he suggests that Sade's inclusion in that series of classics is historically inevitable. And he advances three further arguments. The prospect of "contagion" from an author like Sade is treated as belonging to the realm of "phantasm." Second, the rehabilitation of Sade is presented as an effective means of combating the bourgeoisie with its false principles of moral hygiene. And third, he maintains that Sade offers no greater a moral danger than the descriptions of torture in the lives of martyred saints.

Delon's case for Sade does not survive close examination. The "irresistible evolution" that has supposedly made him a classic really designates the abdication of responsibility by critics who have failed to oppose the shift. The danger of contagion, of effects on the young and the violence-prone, cannot be dismissed with a sneer word. To appeal to an obsolete prejudice against the class that founded our institutions of justice and democracy reveals a singular political naïveté—or bad faith. And Delon (like Paulhan) has exploited and distorted superficial similarities by placing descriptions *condemning* the torture of Christian martyrs in the same category as Sade's *glorification* of torture for sexual gratification.

I have argued that we should not burn Sade, and that we should not glorify him as a new classic of revolutionary moral liberation. What, then, shall we do with him? In order to answer wisely, we must keep in mind a number of basic considerations. We rely on several interlocking institutions—familial, educational, civic, moral, and intellectual—to exercise some control over inexpungible human selfishness and malevolence. We are living out a wager that the freedoms we have won, or have granted ourselves, over the past four centuries have made these institutions more secure. At times, they look more precarious. In either case, each child must learn afresh which qualities of human nature a culture wishes to nurture and which to constrain. Until that process of socialization is well advanced, dangerous and destructive ideas should be admitted with care into the child's environment. C. S. Lewis is eloquent on the point—for both children and adults.

> *That elementary rectitude of human response, at which we are so ready to fling the unkind epithets of "stock," "crude," "bourgeois," and "conventional," so far from being "given" is a delicate balance of trained habits, laboriously acquired and easily lost, on the maintenance of which depend both our virtues and our pleasures and even, perhaps, the survival of the species. . . . When poisons become fashionable, they do not cease to kill.*
>
> (*A Preface to "Paradise Lost,"* Chapter IV)

The world teems with salutary influences and with poisonous influences. The critic's minimum responsibility is to recognize writings for what they are and to puncture false claims.

7. Truth in Labeling

The 1965 Grove Press edition of *Justine and Other Writings* presents Sade for the first time in English as an openly published mainstream writer. The lengthy front matter opens with a Translators' Foreword and a Publisher's Preface. These two critical pieces tell us how Sade has been labeled on the package for thirty years and will be for many years to come.

The publisher (Barney Rosset did not sign his preface) treads softly and cites many of the critics I have discussed who seek to rehabilitate Sade. At the end, he addresses the larger question. "What is strange, and worth investigating" in this "writhing, insensate universe at the pole opposite Gethsemane and Golgotha?" Rosset provides an answer that Milton and Baudelaire could have followed easily.

> *To profit from that extraordinary vision . . . we do not have to subscribe to it. But if we ignore it, we do so at our own risk. For to ignore Sade is to choose not to know part of ourselves, that inviolable part which lurks within each of us and which, eluding the light of reason, can, we have learned in this century, establish absolute evil as a rule of conduct and threaten to destroy the world.*

A few lines later, Rosset repeats the argument. Twenty years after Hitler, Sade's works will "serve to remind us . . . of the absolute evil of which man is capable." Rosset's logic is essentially consistent: The stronger the vaccine, the surer the immunization. He fails to recognize that beyond a certain point of virulence, a vaccine can become a means of infection for certain immune systems. But for Rosset, Sade remains a negative object lesson.

The translators, Richard Seaver and Austryn Wainhouse, take a very different approach. After referring in their first paragraph to Sade's "immense and incomparable literary achievement," they grant that Sade wished for the status of an unknown author with an underground influence. For his "secrets cannot bear disclosure" to the normal commonsense reader. To a reasonable man, Sade "resembles nothing so much as death." But for certain readers Sade's secrets take another course and reach a darker level of response.

> *However firmly* [the reader] *be established in the normality that makes everyday life possible, still more firmly established in him and infinitely more deeply—in the farther reaches of his inalienable self, in his instincts, his dreams, his incoercible desires—the impossible dwells, a sovereign in hiding. What Sade has to say to us—and what we as normal social beings cannot heed or even hear—already exists within us, like a resonance, a forgotten truth, or like the divine promise whose fulfillment is finally the most solemn concern of our human existence.*

Seaver and Wainhouse have here described vividly the subtle way in which Sade's scenes and ideas infiltrate the moral substance of some readers and animate a tendency hidden deep within us. The metaphors and rhetoric of the quoted passage suggest the force of forbidden knowledge conveyed by esoteric writing only to an elect.

In labeling that forbidden knowledge, the translators assign it strong positive value: "a sovereign in hiding," "a forgotten truth," fulfillment of a "divine promise." Their own publisher used more explicit terms: "an incipient terrorist," "the Satanic strain," and "absolute evil." The translators have transformed a negative object lesson into exemplary hero. In the next paragraph, they declare: "It is not our intention to enter any special plea for Sade." But they have already done exactly that by their evaluation of the fundamental nature of Sade's universe. The uninitiated reader of Sade will find these claims on the second page of a foreword introducing "one of our civilization's treasures."

The confusions that emerge from these introductory pages lead me to lay out in schematic form the situation into which Sade and his supporters lead us. The categories I shall set up inevitably eliminate the middle ground many of us would wish to explore and possibly to defend. But clear distinctions serve us best here.

Since Plato and Aristotle, discussions of crime and evil in art can be reduced to two positions: the theory of infection or corruption and the theory of catharsis or purgation.* In a canny examination of the question in the opening pages of *Art and Anarchy*, Edgar

---

*A succinct statement of the opposition appears in the stanza from the sixteenth-century poet Agrippa d'Aubigné that Baudelaire chose as epigraph for *The Flowers of Evil*:

Wind distills a terse version: "Art has the power to intensify (not just to purge) emotions."

As I have already shown, Sade claimed sometimes (in anticipation of the censor) that he wrote about vice and evil in order to cure us of them; and at other times, in order to win readers to them. Modern critics engaged in rehabilitating Sade have laid out a third position: He produced harmless word structures without (intentional) moral dimension, writings toward which we should have an entirely aesthetic response. By combining the above positions, I obtain an outline whose seven ways of reading Sade are not put forward as exhaustive.

A. *Sade is free of moral intention, a wordsmith who produced "a mere combination of texts"* (Barthes).
   1. Sade has no moral effect; purely aesthetic status.
   2. Sade brings (unintentional) moral edification (catharsis).
   3. Sade brings (unintentional) moral corruption (infection).

B. *Sade carries a strong moral component and knew what he was doing.*
   4. Sade is trying to cure us & succeeds (catharsis).
   5. Sade is trying to cure us & fails (infection).
   6. Sade is trying to corrupt us & succeeds (infection).
   7. Sade is trying to corrupt us & fails (catharsis).

Of the seven ways of looking at Sade, the first strikes me as patently untenable. Those who defend it are usually trying to elevate

---

On dit qu'il faut couler les exécrables choses
Dans le puits de l'oubli ou au sépulcre encloses,
Et que par les escrits le mal ressuscité
Infectera les moeurs de la postérité;
Mais le vice n'a point pour mère la science,
Et la vertu n'est pas fille de l'ignorance.

Such filth should be disposed of, men will say,
Nor be allowed to fester and decay,
For, once put into words, rank things may bloom
That send whole generations to their doom;
But knowledge never yet gave birth to Vice,
Nor Virtue looked to ignorance for advice.

(TR. WALTER MARTIN)

Should we bury evil? Or try to know its ways?

Sade to the category of great literature on partial or false grounds. Both Sade's temperament and the psychological stresses under which he lived make it impossible to describe with certainty his intentions in writing what he did. Considering the highly conflicting evidence, it is fully conceivable that he was engaged in an immense wager, a bet with himself that he could wreak his revenge through his writings alone and destroy the society that had deprived him of freedom for half of his adult life.

The translators and publisher of the Grove edition have the sense to acknowledge the moral dimension of Sade's writings. The "sovereign in hiding" to whom Sade presumably speaks in each of us suggests the bringing to life—a little like Frankenstein's monster—of a latent superman in us who is bent on carrying out the exploits described by Sade. The translators seem to mean that for a few great souls Sade will bring immense liberation (item 6), and that other ordinary mortals will draw back in horror (item 7). By calling Sade "one of civilization's treasures," they appear to applaud the destructive element in his work. The publisher, on the other hand, recognizing the evil potential of Sade's imagination, describes it as "surreal rather than real" and values his writings because they will have a healthy effect—whether according to item 4 or item 7 is not clear. Possibly because of the sixties euphoria in the midst of which they were written, both the Foreword and the Preface mock Puritan and Victorian morality as hypocritical and imply that Sade brings a healthy salvation from it.

In the face of these conflicting claims, how can we hope to get Sade right? Above all, how can we estimate the effect on readers of a content and a style with which we have now gained a little familiarity? The arguments I offer now do not close the case. They are intended to turn, even to reverse, the course of current opinion.

I shall take the most difficult matter first: empirical evidence of social harm from writings like those of Sade. The two cases I have examined, the Moors murders and Ted Bundy, could lead us into a vast and inconclusive literature on the sources of violently criminal behavior and on the effects of pornographic and obscene materials on various groups. Instead, I shall focus on two U.S. government reports on pornography. Each has come under violent attack. Each, in my opinion, deserves support. The *Report of the Commission on Obscenity and Pornography* (1970) reaches this conclusion. "Empirical research designed to clarify the question has found no evidence to date that exposure to explicit sexual materials plays a significant role in the causation of delinquent or criminal behavior among youth or adults" (32). The 1970 report favors regulating such materials only for minors.

The *Final Report* by the Attorney General's Commission on Pornography (1986) comes to a different but not opposed conclusion. "Substantial exposure to sexually violent materials as described here bears a causal relationship to antisocial acts of sexual violence and, for some subgroups, to unlawful acts of sexual violence" (326). The 1986 report accepts the earlier negative findings about sexually explicit materials and concentrates its investigations and its findings on sexually *violent* (Class I) materials. They do have socially significant effects. Subsequent discussions have made two crucial points.* It is the element of violence, particularly as associated with and supported by sexual activity, whether explicit or not, that has deleterious social effects. And second, any reference to "cause" in this context means not necessary effects on all individuals exposed to sexually violent materials but on a probable small percentage. This is the kind of statistical link that leads us to call smoking a cause of lung cancer, drinking a cause of automobile accidents, and seat belts a cause of decline in traffic deaths. The 1986 report, while acknowledging a correlation particularly between Class I sexually violent materials in television and movies and sexual violence in society, nevertheless calls for no change in current federal obscenity law under *Miller.*

I go over this ground in order to point out that for the 1970 report, Sade's writings belong to a category of sexually explicit materials that hold no danger to society, whereas for the 1986 report, his writings fall into the most extreme area of Class I, the only category cited as a possible cause of sexual violence in those exposed to it. As other writers and government reports have pointed out, this matter concerns not only First Amendment rights of protection of speech but also considerations of public health and the fitness of the environment. I would say that the 1986 report allows

---

*For a comprehensive survey of the debate with exhaustive references to current research, see Frederick Schauer, "Causation Theory and the Causes of Sexual Violence," *American Bar Foundation Research Journal,* 1987.

us to get Sade right, to place him in the proper category, without calling for an auto-da-fé.

In referring to the potentially criminal effects that violently sexual materials may have on "some subgroups," the 1986 report is acknowledging that most scientific and legal discussions of the effects of pornography and obscenity limit themselves to effects on "nonpredisposed normals"—that is, the average person. But the woods and the woodwork are full of unpredictable temperaments predisposed in many directions. The word *subgroups* opens a way back to the Hicklin rule of 1867. A British court decided that a book entitled *The Confessional Unmasked* "tended to deprave and corrupt those whose minds are open to such moral influences." A more recent American case in the 1950s concerns seventeen lower court decisions against one Winters, a distributor of "adult" comic books in New York City sold mostly to juveniles. The New York Court of Appeals found that "collections of pictures or stories of criminal deeds of bloodlust or lust unquestionably can be so massed as to become vehicles for inciting violent crimes." When the U.S. Supreme Court overruled that finding as vague and unclear, Justice Frankfurter entered an eloquent dissent. "It would be sheer dogmatism . . . to deny to the New York legislature the right to believe that the intent of the type of publication it has proscribed is to cater to morbid and immature minds."

Given the incidence of violence and sex crimes in our society, we would do well to consider the effects of Class I materials not only on "nonpredisposed normals" but also on "morbid and immature minds."* There is a cost-benefit analysis to perform here.

---

*This discrimination among different segments of the public was explicitly—but hypocritically—recognized at the 1956 Paris trial of the Editions J. J. Pauvert for "*outrage aux moeurs*" in publishing Sade's four principal novels. Every witness for the defense testified to the "importance" of Sade's works and went on to insist on the need to restrict the circulation of the four works. Pauvert himself favored limiting the edition by setting a high price and not displaying it in shop windows. These books were to be published for scholars and intellectuals only, and Sade's principal champion, Georges Bataille, was even more restrictive. The reading of Sade "can be only on a reserved basis. I am a librarian." He stipulated that certain formalities should be observed, such as authorization by the head librarian (see *L'affaire Sade*). Judging by subsequent events, we can surmise that Pauvert and his friends were speaking only to persuade the judge. They succeeded. The proposed restrictions disappeared in a few years. Today, the four contested works are widely available in inexpensive editions prominently exhibited in bookstores in

We need to weigh the advantage of free speech and the unimpeded circulation of ideas against the advantage of a balanced environment for the young to grow up in and for the mentally unstable to survive in without doing harm to themselves and others. Stated differently, we need to weigh the benefits of the alleged safety-valve effect of Class I materials on some persons against the danger of such materials affecting subgroups in such a way as to cause antisocial acts. Ivan Karamazov saw the dilemma vividly. "What price will we pay to prevent the torture of one helpless child?" It behooves us to understand that Dostoyevsky is not being sentimental. The question is utterly realistic.

As I have already argued, the burden of proof falls as much on liberals to show that no social damage results from the general availability of Class I materials as on conservatives to show that such damage does or may occur.

Let me return to the medical analogy of infection and immunity. The human immune system functions in complex ways we do not fully understand to protect us from bacterial and viral disease. Its many parts, from thymus gland to T cells, distinguish vital elements in the body that constitute a "self" to be sheltered from alien and dangerous elements marked for attack, and also from those that fall between and can be tolerated as part of a dynamic immunological environment. The proper functioning of this system, supplemented by many others that make up our bodies, leads to a relative state called health—the capacity to recognize and resist encroaching disease. We must marvel at the resourcefulness and sensitivity of our immune system, which seeks not a disease-free environment but one that does not overwhelm our resistances. Health in these circumstances can be described as homeostasis, a kind of steady state for the living self marked off clearly from the contingencies of the external environment. The worst blunder that this finely tuned system can make is to fail to recognize itself and to attack its own cells. (See Tauber, *The Immune Self.*)

The way a child is raised and educated influences its moral immune system, which allows it to participate in the mutual pursuit

---

most countries of the West. The blurbs on the covers lead a reader to believe he or she is buying great literature that makes a fundamental contribution to philosophy and morality.

of happiness within a culture. As a child can be given vaccinations at predetermined ages to activate its immune system against polio, for example, so certain spectacles and stories introduce a child to aspects of violence and evil to which it can develop a resistance. Or it may become infected. In advanced liberal societies, both our physiological and our moral immune systems have been subjected to enormous pressures. In the name of free speech we may defend practices like indecency, desecration, and hate speech while, at the same time, fearing their effect on the community. And we watch the all-pervasive phenomenon of the media feed these practices directly into our moral immune system. Surrounded by these pressures, we are still groping for some kind of commonsense oversight to limit the basest commercial appeals to violent and prurient interest.

Within this broad cultural scene, the writings of the Marquis de Sade represent a small yet symptomatic episode. For in what appears to be the most unequivocal case of Class I materials—extreme violence and mayhem endorsed in close association with utterly explicit sexual excesses and perversions—some of our best-trained minds have argued successfully that we should classify these works as literary and philosophical masterpieces deserving wide circulation. Such a failure of truth in labeling recalls the story of the emperor's new clothes. But the Sade case is infinitely more serious than mere nakedness. Can our perception of what is really at stake in his writings have gone so far astray? Are we dealing here with an advanced form of naïveté, or with calculated cultural nihilism?

Of course one test case lies immediately under our noses. In order to write this chapter, I spent several hours a day for over four months reading Sade's works. And for several years before and after that period, I thought about his case, read other materials on obscenity and pornography, and contemplated the reality of having a shelf of such books in my library and my household. What have been the effects on me of extended exposure to civilized taboo? Do I qualify as a "nonpredisposed normal" or as a subject with a morbid and impressionable mind? Would I counsel such a course of study to others—to moral philosophers, to criminologists, to college students still in their teens? (Most of the above have already explored the area without prompting from me.)

At age twenty-three, I first read a small sample of Sade's works in a public reading room overseen by an elderly woman reference librarian while I was working on an undergraduate thesis on Apollinaire. Almost forty years elapsed before I read him systematically in editions purchased openly in France and the United States. Both times, the experience of entering Sade's universe bore no resemblance to my knowledge of love, passion, copulation, debauchery, and tenderness. Sade's narrative plunged me into feelings and reactions associated with two vividly familiar experiences of a very different kind: witnessing a major surgical operation for the first few times and participating in wartime combat fighting at close range. In watching surgery, from the moment of the first incision into the body, one fights giddily for detachment, distance, and rational justification for so unnatural an act. In combat, one may rapidly yield either to numbing fear or to surging impulses of aggression and bloodlust. In both cases, the universe has been turned upside down. What was wrong is now right in the name of some higher cause, medical or military. Insofar as I can reconstitute those distressing early experiences of reading Sade before it became a chore to continue, I sought both to withdraw and to succumb. Revulsion accompanied arousal to produce a kind of visceral trembling that resembles stage fright. Cold blood and hot blood mingled and fought one another in a state of tense paralysis. In spite of its intensity, this condition was the opposite of life, or health, or pleasure.

No, I do not believe I have been harmed by this experience. But I can imagine it having such an effect. Conversations with friends and associates inform me that their reactions vary a great deal and that they find themselves equally unharmed. Perhaps we are all laying claim, as Descartes suggests, to an adequate portion of common sense. Most of us also believe that the effects on some troubled minds could be far stronger, more lasting, possibly dangerous.

To know the world and the human beings in it, one does not have to visit every country, let alone the North Pole. Sade's writings represent perhaps not the moral North Pole itself but so vivid an account of it that a few unhinged individuals will be induced to attempt the expedition and to visit the place. The social crimes of the Moors murders and of Ted Bundy's case furnish examples of such a response. Could they have been avoided? We do not know. But one of the least advisable measures to that end is to classify Sade's works as great literature.

Sade frequently cited originality as his claim to immortality. But he had a predecessor in this raid on fame through infamy. It is the nearly forgotten Greek figure Herostratus. According to legend, this undistinguished and profoundly thwarted citizen of Ephesus conceived the idea of burning down his city's temple of Artemis with its fine library in order to create instant fame for himself and assure the survival of his name in history. Out of calculated self-aggrandizement, Herostratus committed an act of cultural arson, causing the destruction of genuinely valuable artifacts and probably of human lives.

What shall we do with such a story? For if we perpetuate the story as a negative example, we are also perpetuating the success of his crime. And if we try to suppress the story because of possible misinterpretation and deleterious consequences, we are establishing limits on knowledge of history and losing a fable. What we can and should properly protest is not the existence of the Herostratus parable but any interpretation of it, particularly for young and unformed minds, as recording a model deed of originality, courage, and human liberation. Herostratus could not project his imagination beyond furthering his own selfish interests by devastating the interests of others. Like Sade, he was engaged in the destruction of the very history in which he wished to survive.

The divine marquis represents forbidden knowledge that we may not forbid. Consequently, we should label his writings carefully: potential poison, polluting to our moral and intellectual environment.

Chapter VIII

THE SPHINX AND THE UNICORN

■

I know only that I do not know.

—Socrates

The most incomprehensible thing about the universe is that it is comprehensible.

—Einstein

A contradiction or paradox lies buried in the title of this book, *Forbidden Knowledge.* If we are familiar enough with any entity or domain to call the result "knowledge," then we already know too much about it to apply the adjective *forbidden.* The taboo or prohibition has already been broken, the obstacle or risk overcome. The only true items in the class, then, would be forms of knowledge still unlocated, unnamed, unexplored, possibly closed to us. "How," Meno asks Socrates, "will you look for something when you don't in the least know what it is?" These paradoxes do not disqualify the phrase "forbidden knowledge." On the contrary, the phrase remains with us and carries meaning by its long association with particular stories and case histories.

A reader seeking a breakdown of forbidden knowledge into categories designed to contain the materials of the previous chapters should turn now to Appendix I. The six categories I propose there lend some order to the variety of stories and case histories I have referred to. And the categories provide the closest approach I make to a theory of forbidden knowledge.

In this chapter, I pursue toward its outcome the inquiry with which I began: Are there things we cannot or should not know?

1. WHAT WE OUGHT NOT TO KNOW: THE INSTITUTIONS OF SCIENCE AND ART

In the long perspective of four thousand years, the Western world has discovered or invented only two master plots, two narratives of high explanatory power. They affect every aspect of human life today. The amalgam of the Greco-Roman heritage with Jewish and Christian traditions produced a culture in which the naked rule of status and power began to yield to justice under law, the dignity of all persons under God, and a morality of altruism. The sequence of Hebrew covenants followed by the New Testament Redemption story offers an account of things bestowed on us from on high by a single God. That religious ethos, though reformed and attacked for the last five centuries, had no full-fledged competitor until the elaboration of Darwinian evolution in the middle of the nineteenth century. Renaissance secular humanism and Enlightenment reason opened the way for evolutionary theory without themselves establishing a complete and competing account of things. The new story of life emerging uncreated out of the primal slime and finding its upward way by natural elimination (misleadingly called "natural selection") has partially displaced the old story without destroying its teachings and ideals. Since the Renaissance and the Enlightenment, many citizens of the West have composed or compartmentalized themselves in such a way as to accommodate both stories.

We have barely begun the momentous struggle between these two master plots, Christian and Darwinian.* The widely admired writings of Nietzsche, intent on the destruction of Christian ideals, would return us to an atavistic morality of power, status, and cruelty exercised by noble masters. Nietzsche had drawn deeply from Dar-

*Chapter VI may imply that technology, closely linked to science and commerce, confronts us with a third master plot. But by itself, technology has no story. It relies on progress as its sustaining myth.

win. A succinct and unflinching answer to Nietzsche arose out of Martin Luther King, Jr.'s resolve to protect the civil rights struggle from the forces of radical black violence. In "Where Do We Go from Here?"—his 1967 presidential address to the Southern Christian Leadership Conference—King picks out as one of the great errors in history the interpretation of power and love as polar opposites and the association of power with violence. King cut to the core of the matter with a no-nonsense simplification.

> *It was this misinterpretation that caused Nietzsche, who was a philosopher of the will to power, to reject the Christian concept of love. It was this same misinterpretation which induced Christian theologians to reject Nietzschean philosophy of the will to power in the name of the Christian idea of love. Now, we've got to get this thing right. What is needed is a realization that power without love is reckless and abusive, and love without power is sentimental and anemic. Power at its best is love implementing the demands of justice.*
>
> (*A Testament of Hope*, 247)

King was not just playing games with the words *love* and *power.* He was reaching back to a series of his own earlier readings (above all, in Paul Tillich) and writings and to his experience as intellectual and tactical leader of the civil rights movement. "To get this thing right" meant to King an appeal to a long-meditated and carefully defined philosophic position: the philosophy of nonviolence. In such talks as "The Power of Nonviolence," given in 1958 for the YW-YMCA in Berkeley, California, King explained the intellectual conviction, personal discipline, regular training, physical courage, Gandhian *Satyagraha*, and Christian agape needed to carry out nonviolent resistance. And the struggle he led was not between two peoples or races, but "between justice and injustice, between the forces of light and the forces of darkness."

Nietzsche's "master morality" based on "the will to power" has powerful affinities with social Darwinism. It appeals to convictions and disciplines that turn their back on any "slave morality" of love and pity. These two prophets, Nietzsche and King, confront us with a continuing struggle between power and justice that no thinking person can responsibly turn away from.

The notion of forbidden knowledge has emerged out of both these master plots, Judeo-Christian and Darwinian. A large collection of familiar myths, including Prometheus and Pandora, Adam and Eve, and most of the other stories I have been examining, traces the development of forbidden knowledge in the Greco-Roman and Hebrew traditions and on into the Christian heritage. It is more surprising to discover that the notion of forbidden knowledge has also appeared in the far more recent Darwinian dispensation. I am referring not only to episodes like the restrictions placed on recombinant DNA research in the 1970s by the investigators themselves but also to reservations and second thoughts expressed by some of the most dedicated spokesmen for the Darwinian view of existence.

For example, bringing together a mass of new work in ethnology, comparative psychology, and population biology, Edward O. Wilson published in 1975 an immense tome, *Sociobiology: The New Synthesis.* It begins by dismissing "solipsist consciousness" and by affirming natural selection as the all-powerful principle among living things, particularly in social behavior. Then, after five hundred double-column encyclopedic pages mostly on animals and social insects, Wilson comes back to human beings and the dilemma posed by our expanding knowledge of our own "machinery." Sociobiology will "cannibalize" psychology, policy, ethics, and even molecular biology. Accordingly, Wilson believes that scientifically planned society is inevitable in the next century.

At this point, something startling happens, like Orpheus looking back. On the last page, Wilson announces calmly that by usurping natural selection, "social control would rob man of his humanity." In the concluding paragraph of the book, he raises, in a sudden lunge, perplexing questions about knowledge and human nature, as if he were appalled by the claims he has made. "To maintain the species indefinitely we are compelled to drive toward total knowledge, right down to the levels of the neuron and gene. When we have progressed enough to explain ourselves in these mechanistic terms, and the social sciences come to full flower, the result might be hard to accept" (575). Wilson's belated apprehensiveness about the consequences of planned intervention in natural selection lead into a final, cryptic quotation from Camus about man becoming an alien "divested of illusions and lights." Wilson cannot retract the book on its last page. But the sense of a mental reservation, of forbidden knowledge, has been wrung out of him by his own argument.

Wilson represents a large number of scientific figures who have expressed hesitations over the most imperialist and reductionist claims of the Darwinian master plot.* The wisest of them state in one way or another that in this domain we may now know too much too soon.

The simplest way to return from these lofty speculations about master plots to the preoccupations of our daily lives is to raise the basic question posed in the Foreword: *Are there things we should not know?* At least four answers—religious, philosophical, historical, and literary—deserve serious attention.

The religions of the West (as well as most Eastern faiths) answer yes, there are things we should not, cannot, need not know. To probe brashly beyond what God has revealed to us and to explore final questions by reason alone will distract us from the responsibility of living our lives according to an established moral code. Faith in a higher being directs us not to undermine his place by seeking Bacon's "proud knowledge" but to approach knowledge as a means of admiring his handiwork and submitting to it. We revere great learning, but it may be of the Devil's party. In the religious tradition, salvation comes from faith, good works, or Providence—not from great knowledge alone. Nicholas of Cusa's "learned ignorance" approaches closer to the religious impulse than does Teilhard de Chardin's optimistic Catholic scientism.

Insofar as they separate themselves from religious faith and control, philosophers (including "natural philosophers," or scientists) have tended to give a negative answer to the question whether there are things we should not know. By definition, philosophers love knowledge *(sophia)* and recognize no external authority to limit their hypotheses and inquiries. A few scientists such as Oppenheimer have muttered darkly about "knowing sin" in their work, and a group of geneticists once imposed a slowdown on themselves in

---

*See the following entries in the Bibliography: Jean-Pierre Changeux, Carl Degler, Troy Duster, Gerald M. Edelman, Gerald Holton, François Jacob, Evelyn Fox Keller, Daniel J. Kevles, Arthur Koestler, R. C. Lewontin, Jacques Monod, James V. Neel, Melvin Konner, and Alfred Tauber.

recombinant DNA research. On the other hand, the philosopher, Nicholas Rescher, in *The Limits of Science* argues for the theoretical *limitlessness* of scientific inquiry. In general, secular philosophy and science have articulated no strong principle that would retard or constrain the practices that have produced what we call "the knowledge explosion." Even ordinary prudence loses ground to ambition, greed, and the sheer momentum of discovery.

There are, however, odd recesses and crooked paths within the philosophical-scientific perspective. From Socrates' irony to Erasmus' comic monologuist, Folly, to Einstein's space elevators and black boxes and "spooky action at a distance," the human mind has at times looked again at human knowledge and seen in it an elaborate cosmic joke. Today, physicists talk emphatically of the "craziness" and "weirdness" of events occurring close to the speed of light and to quantum forces. "If you really believe in quantum mechanics," states the physicist and mathematician Robert Wald, "then you can't take it seriously." Lewis Carroll in his Alice books and Alfred Jarry with his science of 'Pataphysics ("The science of laws governing exceptions") explored the cosmic joke from the literary side.

The same sense of yawning immensity and vanishing tininess, when grafted onto the philosophical tradition of doubt and skepticism, can lead not to laughter but to visceral revulsion from the void emptied of God's presence. "The eternal silence of these infinite spaces frightens me," wrote Pascal, as much a scientist as a believer. In moments of vertigo, the modern scientist-philosopher may behold himself as adrift between the ultimate and opposite constants, c (the speed of light) and h (Planck's quantum). The enormous appeal to intelligent adults not only of Alice but also of Beckett's two abandoned clowns in *Waiting for Godot* arises, I believe, from Didi and Gogo being equally responsive to the cosmic joke and to *horror vacui*. But these probing and crotchety insights into the nature of things do not modify the general answer of philosophy and science to the question "Are there things we should not know?" They answer no.

There is more to learn from history's answer than from the two previous answers. Here much of the work has been done for me by Hans Blumenberg in *The Legitimacy of the Modern Age* (1983). Part III of that scholarly work on the history of ideas devotes two hundred pages to a systematic history of curiosity, the thirst for knowledge, from earliest antiquity. Inevitably and properly, the refrain of Blumenberg's account is furnished by the opening sentence from Aristotle's *Metaphysics,* which I usc as an epigraph for this book. "All men possess by nature a craving for knowledge." And Blumenberg does not fail to cite the two oldest anecdotes on the subject, both of which concern a well.

> *. . . the story about the Thracian maidservant who exercised her wit at the expense of Thales, when he was looking up to study the stars and tumbled down a well. She scoffed at him for being so eager to know what was happening in the sky that he could not see what lay at his feet.*
>
> (Plato, *Theaetetus*)

In Democritus' counterpart tale, an unnamed philosopher is bending over a well to look for the truth. But the truth has withdrawn into the depths of the earth and will not reveal itself. Both parables tell us that curiosity may tempt us away from what is most important: the life that lies immediately in front of us.

Blumenberg traces first how the Greeks and Romans gradually reached a cautious answer to the question of how much we should try to know. The Thales fable suggests that astronomy may be a foolish distraction. By the time Cicero summed everything up in the first century B.C., he could propose a median view that encouraged knowledge of nature—even astronomy—as good training for essential knowledge: practical life, morals, and politics in the largest sense of social responsibility. In *De finibus,* Cicero censured Ulysses' behavior toward the Sirens as motivated by pure greediness for knowledge that distracts him from returning to his duties in his native land.

This reasonable solution had to yield slowly to the doctrine and dogma of the Roman Catholic church, most tersely stated by Tertullian, "After Christ, we have no need of curiosity." One had best tend to one's own salvation. The final truths had been revealed in Scripture. In the early Middle Ages, curiosity was seen as a consequence of acedia—apathy and indifference toward the true purpose of a devout life. But by readmitting Aristotle into their midst,

Aquinas and the scholastic theologians gave secular knowledge and even curiosity a new start. Blumenberg misses Aquinas' pertinent discussion in the *Summa theologica* (II, qq. 166–67) where he distinguishes between *curiosity* and *studiousness.* The former, driven by pride, vanity, the impulse to sin, or superstition, leads us astray. Studiousness, on the other hand, falls under the virtue of temperance and leads to the knowledge of sovereign truth.

In 1336, Petrarch confessed in a famous letter that he climbed Mount Ventoux out of pure curiosity. When he reached the top and looked down like a god upon the world, a deeply medieval revulsion overcame him and prompted him to take out his pocket Augustine. He turned providentially to a passage condemning just this kind of worldly distraction from his devotion to the true faith. But after Copernicus and Galileo and the voyages of discovery, after Bacon, Newton, and Descartes, philosophers found new reasons to legitimize and encourage curiosity and to oppose all restrictions on knowledge.

Present-day historians tend to associate this shift more with the northern Enlightenment than with the earlier Italian Renaissance. But the former would never have occurred without the latter. The most incisive acknowledgment of the shift is Goethe's adoption of Lessing's scheme to transform the heretical and sinning Dr. Faustus into the heroic, striving Faust. Faustian man certifies the secularization of Western culture and its new freedom to explore forbidden knowledge.

Blumenberg's history of "theoretical curiosity" continues. I pause here to take account of what has happened. The remarkable intellectual achievements of a series of individuals during the seventeenth and eighteenth centuries in France, England, Scotland, Germany, and the United States led to the formation of two new institutions. In three centuries, those two institutions have profoundly transformed the fabric of our lives and the forms of our thinking.

On the one hand, we extol science as a collective activity based on experiment, highly perfected instrumentation, and the conventions of the scientific report. Enterprises like the British Royal Society, founded in 1660, and the successive volumes of Diderot's *Encyclopédie* (1751–1772) gave to science the rudiments of independence from church and state. Today, scientists form a clerisy and something approaching an international government of their own, responsible primarily to themselves as guardians of empirical inquiry.

On the other hand, we extol art, no longer as a traditional practice tied closely to notions of craft, the imitation of beautiful forms in nature, and moral utility, but as an individual creative activity springing from original genius, reliant on a disinterested "aesthetic" attitude, and free of social constraints. Lord Shaftesbury in England and Kant's *Critique of Judgment* (1790) in Germany codified these notions during the eighteenth century into a new class of experiences and judgments concerned with "fine art." The most conceptual of philosophers with little experience of the arts, Kant insisted in the opening passage that the beautiful consists in "pure disinterested delight."

Today, we see science and art as essentially opposed activities. It is worth pointing out the close proximity of their origins in "the disinterested attitude." In science, it became the ideal of objectivity and impersonality in the pursuit of empirical truth; in art, it became the aesthetic attitude, art for art's sake, and the separation of art from utility* and morality.† And from the start, both science and art employed the word *experiment* to refer to their new endeavors and products.‡

Furthermore, each of these powerful new institutions to a large extent occupied intellectual and even spiritual terrain that previously had belonged to religion. Science and art became semipriestly vocations holding out the promise of improving the lot of mankind and calling for dedication and faith. T. H. Huxley in his defense of scientific education and Max Weber in his talk to students, "Wissenschaft als Beruf," describe science as a calling that demands sacrifice and discipline. The religious aspirations of art are even more visible in the widespread elevation of the artist to the role of

---

*"Anything useful is ugly," wrote Théophile Gautier in the Preface to *Mademoiselle de Maupin* (1835).

†Poe ranted against "the didactic heresy," an expression adopted by Baudelaire.

‡In the second sentence of the Preface (1800), Wordsworth refers to *Lyrical Ballads* as an "experiment." Whitman called *Leaves of Grass* "only a language experiment."

a new priest. The German idealists, including Kant and Hegel, argued that the artist will reestablish our lost communion with the spiritual and the transcendent.*

Thus out of the long history of curiosity and its development into forms of freethinking emerged two modern institutions claiming increasing autonomy from religious and social constraints. Today, relying on principles of experiment, pure research, free speech, artistic license, and academic freedom, science and art can affirm a measure of independence from limitations on ordinary behavior. In extreme instances, each has claimed to be a no-fault activity occupying a morally tax-free zone. In 1994, a gangsta rap performer in New York City named M. C. Pooh appealed to the separation of art from life in order to justify as "art" his own criminal acts and incitements to crime in his performances. Genetic experiments whose outcome might undermine our biological equilibrium more seriously than the atomic bomb are proposed on the principle of pure research. The eternal human trait of curiosity has constituted itself into two powerful institutions. Science and art have enlarged our way of life; in extreme cases, they may now also endanger it.

History, then, leads us back to the question "Are there things we should not know?" History's record suggests that we pay attention to the extended and instructive shift from Cicero's measured encouragement of knowledge, to the restrictions of the medieval world ("After Christ, we have no need of curiosity"), to the increasingly open approach of the modern age. Our passwords today are *experiment*, *originality*, and even *subversion* as embodied in our two established—and sometimes rogue—institutions of science and art.

Sensible interpretation of this history of curiosity concerns us very deeply. I do not believe we should read the account of these developments down to the present exclusively as the record of gradual liberation from superstitious restrictions on human creative and imaginative powers. For the same history furnishes a cautionary tale, telling us that complete liberation from constraints in the arts and sciences may endanger our humanity and the fragile entity we call civilization. Even the most ancient story of starry-eyed Thales

*Paul Bénichou has published a series of volumes devoted to the history of the artist as secular priest.

falling into a well has not lost its point. Our very accomplishments can distract us from seeing accompanying perils. Ulysses' shipwreck while in search of new worlds as imagined by Dante foreshadows *Frankenstein*, unmentioned by Blumenberg but an essential component of his history. The atomic bomb, recombinant DNA, and the Human Genome Project provide further test cases of curiosity cultivated as necessary or self-justifying.

The history of curiosity points to something simple and fundamental. Pascal's insistence on *portée*, or the reach of human faculties to understand the universe, is usually associated with our inability to grasp the infinitely small and the infinitely large. We are not at home at orders of magnitude remote from our own. But *portée* also refers to *time*, to the pace at which we can assimilate new discoveries and innovations in technology and in moral attitudes. In his essay "What is Enlightenment?" Kant cautiously restricts the free exchange of ideas because of his strong sense of their potential effect on society. "A public can only slowly arrive at enlightenment." Furthermore how do we distinguish beneficial change from harmful change? Enlightenment from barbarism? We are aware of what damage too sudden modernization may inflict on a backward society. The Ik tribe in Africa reverted to virtual savagery when deprived of its hunting grounds. Yet we inflict unthinkingly on ourselves influences that subject us to enormous stress. Do we know how to resist TV violence and addictive drugs, particularly when promoted by the profit motive? Do we still have the capacity, like an immune system, to reject forces and practices that damage our collective and individual health? My chapter on the Marquis de Sade points out that violent pornography and obscenity should be seen as problems not exclusively of free speech but equally of public health and safety. Statistical studies show that, like radiation affecting genetic material, violent pornography may act on some unstable temperaments to provoke criminally unsocial behavior. Our sciences and technologies, our arts and media, run far out ahead of many citizens' capacity to adapt to their enticements. Yet those two institutionalized forces now draw our culture into the future as fatefully as Helios' two steeds drew the sun each day across the heavens for the Greeks. I suggest that the lesson of the history of curiosity is one of pace and timing. Can we still control our own velocity? Our rate of change? Are we courting the fate of Icarus?

There is a fine short story by Kipling, "The Eye of Allah," that addresses this challenge through the events of another era. At St. Illod's monastery in thirteenth-century England, lay brother John stands out both as the most prized manuscript illuminator and as a man of immense scientific learning. He also travels periodically to Spain, recently reconquered for Christendom, and brings back a precious cargo of colors for the Scriptorium and drugs for the Infirmary. On his last trip, he found new forms of devils to draw surpassing "our Church-pattern devils."

Abbot Stephen is devoted both to the Church and to worldly knowledge. John attends a dinner given by the Abbot to honor two Rogers: Roger of Salerno, a renowned Italian physician, and the pugnacious scientist-philosopher Roger Bacon. John shows to the guests his grotesque new breed of devils and maintains they are not drug-induced, but drawn from nature. The Abbot then invites John to demonstrate a magnifying device called "the Eye of Allah," after the Moors who discovered it. Looking through the "eye," the company observes horrible lumpish shapes in a drop of stagnant water: life forms, or devils alive in Hell? Roger Bacon becomes highly excited over this "Art optical," which will reveal the truth of the world. But the Abbot states categorically that the Mother Church sees these images as a form of magic. It could send them all to the stake for heresy. After destroying John's magnifying instrument with a hammer-blow, the Abbot pronounces that such a device would "enlighten the world before her time . . . this birth, my sons, is untimely. It will be but the mother of more death, more torture, more division, and greater darkness in this dark age."

In a short story that carries the condensed action and moral freight of a novel, Kipling implies that inappropriate and premature knowledge can do great damage and would not in this case bring the Dark Ages to a beneficial close. The Abbot appears to have the experience and sagacity to make such a decision in a hierarchical system of authority. Without such clear authority today to judge the timeliness of knowledge, we seem powerless to resist any novelty or revolutionary invention or commercially driven temptation that presents itself. Our technology and our freedom have advanced very far. But the restricted pace at which a society can absorb innovation, difficult as it is to ascertain that pace, should exert some braking force on the advanced parties of science and art. Kipling's story submits to history as a form of fate we must not slow or hasten. Before its appointed time, Art optical must be forbidden, for premature knowledge may harbor danger. But who will act as our Abbot or our umpire in an era of competitive expansion and of appetites whetted by seductive advertising?

The historical outlook recommends a certain patience in our exercise of curiosity. But we have in our midst today few agencies to encourage patience and restraint.

The literary answer to the question about things we ought not to know subsumes the three previous responses and includes them in a collection of tales whose assembled wisdom is highly complex. The stories I have treated reveal how often curiosity pushes us toward presumption and hubris and *pleonexia*, and how rarely we take full account of constraints like *portée* and moderate pace. Faustian man overwhelms us on all sides. That appears to be our fate at the close of the second millenium. Nevertheless, Faustian man cannot quash a strain of quietism in some of us that does not need to soar to great heights in order to find full humanity.

2. The Veil of Ignorance and the Flame of Experience

In our search how to conduct our lives, driven on one side by curiosity and constrained on the other by our sense of reach, we have been dealt two wild cards. First, the Wife of Bath effect makes us push against any force that appears to limit our freedom of action. The perverseness of the Wife of Bath effect spreads very far. Any intelligent parent must take account of its sway in bringing up a child. The positive value assigned in some quarters to "transgression" belongs to the same impulse. The "don't fence me in" syndrome both protects us from domination by others and nudges us toward antisocial forms of egoism that may thwart justice and decency. This dilemma carries us back very close to the Nietzsche-King debate on power versus love, evolution versus Christianity.

The other wild card is far less familiar. I have discussed a form of it in the third section of Chapter I under the term "fog of uncertainty." Nicholas Rescher's phrase refers to a fundamental condition: What we think of as our humanity entails not having complete insight into the motives and intentions of other people and even into our own. Omniscience would confound us utterly. If we had such devastating knowledge, we would be either gods (as Satan croons to Eve in his temptation speech) or puppets (as Nathanael imagines himself to be in "The Sandman"). "The veil of ignorance" (as I prefer to call it) defines our humanity in both senses of the word *define*: to describe and to limit.*

That blur or cloud lodged at the center of our being, the knowledge forbidden to us in the very midst of our existence, lies at the heart of Faust's lament at the beginning of Goethe's play.

All I see is that we cannot know!
This burns my heart.

(364–65)

In the first part of the drama, Faust tries to surpass his limitations through the magic powers obtained from Mephistopheles. The effort leads to disaster in the Gretchen episode. In the first scene of Part II, Faust wakes up alone with a new lease on life and a new willingness to lower his sights. For after being blinded by the rising sun, he turns back to the earth and seeks shelter behind "the most youthful of veils." The image that immediately follows of a rainbow in the mist informs us (through association with the 1784 poem "*Zueignung*," "Dedication") that Faust-Goethe here finds protection from blinding Truth behind the veil of poetry. Absolute Truth paralyzes. The intermediate realm of poetry shields one from it and allows freedom of movement. Faust now galavants through five acts of wild adventures worthy of a spaghetti Western. The

*"The veil of ignorance" here refers to a real and fundamental aspect of our human condition, both a limit and a safeguard. The political philosopher John Rawls uses the same term to designate a very different notion: a hypothetical situation (ignorance about one's own social status) designed to promote fairness in reaching agreements with other members of a community. "The principles of justice are chosen behind a veil of ignorance," Rawls writes in the opening chapter of *A Theory of Justice*. He means a state of affairs imagined or artificially induced in order to attain a specific social goal. "The veil of ignorance" as I use it means a condition we cannot escape. It bears comparison to Plato's analogy of the cave.

veil motif at the opening suggests that through all his escapades in Part II, Faust knows that he does not know the Truth.*

Without Goethe's vivid metaphors, Kant reaches much the same conclusion at the end of a passage on how far our cognitive faculties can understand nature. Kant's version of the veil of ignorance does not lend itself to any peeking.

> *It is, I mean, quite certain that we can never get a sufficient knowledge of organized beings and their inner possibility, much less get an explanation of them, by looking merely to mechanical principles of nature. Indeed, so certain is it, that we may confidently assert that it is absurd for men even to entertain any thought of doing so or to hope that maybe another Newton may some day arise, to make intelligible to us even the genesis of but a blade of grass from natural laws that no design has ordered. Such insight we must absolutely deny to mankind.*
>
> (*Critique of Judgment*, II, Section 14)

Molecular biologists and geneticists may laugh at Kant's affirmation of ignorance. But they have not yet solved the dilemmas of reductionism and infinite regress.

Poets are drawn powerfully to the veil of ignorance. A. E. (George Russell) composed a four-quatrain poem entitled "Truth" and ended it on the motif of inaccessibility.

> *And only the teaching*
> *That never was spoken*
> *Is worthy thy reaching,*
> *The fountain unbroken.*

The American poet Randall Jarrell finds the mortal condition so elusive that he can only grope toward his yearnings and resign himself to missing them.

---

*Both the first scene of *Faust II* and "Dedication" echo motifs of ascent (wings, heights) and light (illumination, blindness) from Plato's cave parable in *Republic* VII and from Dante's *Paradiso*. Brittain Smith pointed me toward the veil metaphor in Goethe.

If I can think of it, it isn't what I want.

("SICK CHILD")

This is more than just another Irish bull.

The ultimate truth must remain an ineffable mystery. Novelists, too, confront this circumstance. In order not to disturb the object of our knowledge, we may have to turn away from it, as Alyosha in *The Brothers Karamazov* cannot raise his eyes to look at the figure of Jesus in the conversion dream ("Cana of Galilee"). Near the end of *The Mill on the Floss,* George Eliot as narrator allows herself a puzzling yet probing sentence (quoted earlier) about Maggie's moral dilemma: whether she should seize for herself a reward that will make those closest to her miserable. "The great problem of the shifting relation between passion and duty is clear to no man who is capable of apprehending it" (see page 118). Eliot's sentence read in context means that Maggie's situation escapes us if we try to reduce it to moral maxims. That kind of "apprehension" cannot match the "minute discrimination" of particulars enacted by a narrative. But Eliot is also implying that even in a story, "the shifting relation between passion and duty" lowers around us the veil of ignorance that belongs to life itself. Don't try to understand everything. At the end of Chapter III, I call attention to the Hamlet motif in *Faust,* the way Faust's self-awareness impedes the kind of action he seeks. We have approached very close to my fourth category of fragile knowledge. (See Appendix I.)*

These examples point toward an extreme form of forbidden knowledge that I shall call "consciousness kills." At the moment of knowing, we are unable to know. The nature of time combines with the nature of consciousness to produce a succession of evanescent moments that forever escape us. "The specious present," William James called it. The phenomenon was not lost on Nietz-

---

*In *Emile,* Rousseau's Savoyard Priest opens his sixty-page "Profession of Faith" with an extended disquisition on the veil of ignorance. Not only do we remain ignorant of the universe; "we do not know ourselves, we know neither our nature nor the spirit that moves us; we scarcely know whether man is one or many; we are surrounded by impenetrable mysteries." My epigraph from Goethe acknowledges a complementary condition: "*Individuum est ineffabile*"—a bleak statement, also respectful.

sche. "Every living thing needs a surrounding atmosphere, a shrouding vapor *[Dunstkreis]* of mystery." Nietzsche repeats this insight over and over in his early essay on "History in the Service and Disservice of Life." Toward the end of it, *Historie* assumes for him the meaning of consciousness itself, of self-awareness as an interference with active living. Proust takes the same vapor analogy from schoolroom physics and develops it more vividly than Nietzsche. In the remarkable passage I quoted on page 160, Proust offers a searing image of conscious knowledge as self-defeating, self-consuming. The inner sanctum, if we ever reach it, has already been emptied by our own clanking approach. We would do better to keep a discreet distance. The unexamined life may not be worth living, as Socrates taught. Yes, and the overexamined life may bring confusion and paralysis if one tries to pierce the veil of ignorance.

My last quotation comes from the historian and philosopher of science Helen Fox Keller, in a discussion of how genetic science claims to reduce or even eliminate "the locus of freedom" that might permit us individual choice. Whether or not she has Socrates in mind, I believe Keller is highly serious. ". . . the very possibility of choice depends on a residual domain of agency that can remain free only to the extent that it remains unexamined" ("Nature, Nurture, and the Human Genome Project"). Some readers will find that sentence totally opaque. I hear a scientist cautioning herself and us about the encroachments of knowledge and consciousness beyond the veil of ignorance that surrounds—and constitutes—the kernel of our humanity. Don't lift that veil unthinkingly.*

Educated to value enterprise and originality in all endeavors, we rarely have the patience to tolerate such intellectual restraint. In our eagerness to sweep away secrets, we leave no privacy unentered, either in persons or in nature itself. The effects of this impatience in scientific research and in social analysis are not hard to observe. Somewhat less evident are the moral effects on our personal lives. I have taken care to point out how, at key moments in

---

*For nearly a century, Freudian psychoanalysis persuaded us that the unconscious acts essentially to repress memories that we need to recover for therapeutic reasons. Revisionist schools of psychology now attach productive rather than repressive functions to the unconscious. Many of our capacities to think and to behave creatively may depend on the existence of an automatic self, over which we do not exercise conscious, voluntary control. (See Jonathan Miller.)

the Ulysses sequence in Dante's *Inferno* and in Adam and Eve's self-justification in *Paradise Lost*, the word *experience* takes a leading role to designate a self-validating form of action. Life as experience or experiment—this vision links Eve and Adam to Psyche, links Ulysses to Dante, Dr. Faust to Dr. Frankenstein, Don Quixote and Don Juan to Dr. Jekyll. At the furthest extreme, Sade's fiendish heroes carry out their research projects in the pleasure of sheer malevolence. Among the stories I have discussed, only the Princesse de Clèves and Emily Dickinson's "Veil" narrator resist the temptation of the experimental life. The impulse toward experience surrounds us like the air we breathe; it expresses our acceptance of the Wife of Bath effect and our rejection of the veil of ignorance.

The classic epic and the modern novel have unstintingly celebrated the impulse toward experience. Yet the most concentrated treatment of the impulse comes out of the philosophical-moral musings of a retiring Oxford classics scholar in the late nineteenth century. Walter Pater's "Conclusion" in *Studies in the History of the Renaissance* (1873) reads so much like an impassioned manifesto of modern aesthetic hedonism that Pater himself chose to remove it from the second edition. "It might possibly mislead some of those young men into whose hands it might fall," Pater explained. Oscar Wilde knew the "Conclusion" by heart and called it "my golden book." Here lies the Siren song of our era and the voice of an unrepentant Faust.

> *Not the fruit of experience, but experience itself, is the end. A counted number of pulses only is given to us of a variegated, dramatic life. How may we see in them all that is to be seen in them by the finest senses? How can we pass most swiftly from point to point, and be present always at the focus where the greatest number of vital forces unite in their purest energy?*
>
> *To burn always with this hard, gemlike flame, to maintain this ecstasy, is success in life. Failure is to form habits.*

A candle flame, a streaming, consuming flux that somehow maintains constancy of form, offers the perfect image for intensity of pure experience, its "splendour," as Pater writes earlier. In what follows, as in the first sentence of the above quotation, it is difficult to detect how sheer experience attains to any wisdom beyond itself. In his closing lines, Pater tries to convert this hedonism into aesthetic epicureanism and adopts the phrase "art for art's sake." But the fleeting intensity of experience wins out over the permanence of art.

> *We are all* condamnés *as Victor Hugo says . . . we have an interval, and then our place knows us no more. Some spend this interval in listlessness, some in high passions, the wisest, in art and song. For our one chance lies in expanding that interval, in getting as many pulsations as possible into the given time. Passions may give us this quickened sense of life, ecstasy and sorrow of love. Only, be sure it is passion—that it does yield you this fruit of a quickened, multiplied consciousness. Of this wisdom, the poetic passion, the desire for beauty, the love of art for art's sake, has most; for art comes to you professing frankly to give nothing but the highest quality to your moments as they pass, and simply for those moments' sake.*

This alluring voice makes Goethe's Mephistopheles sound like the strutting clown he really is. Tepid in most of his other writings, Pater here finds the subtle appeal of Satan seducing Eve in *Paradise Lost.* Pater's curiously explicit phrasing was not to be surpassed in our time by Alfred Kinsey counting orgasms and Michel Foucault sacrificing himself to *l'expérience limite.* A strong tradition within modern literature has led us toward this worship of pure experience without restraint of any kind.* However, what looks to Pater like "a hard, gemlike flame" can escalate into forms of violence and destruction in order to sustain that fleeting intensity. In the most

---

*We can locate this tradition, for example, in William Blake's Proverbs of Hell. "The road of excess leads to the palace of wisdom." We shall never know how much diabolical irony lurks in those words. Blake's following proverb extends the claim. "Prudence is a rich ugly old maid courted by Incapacity." Blake's apparent appeal to license and daring leads us back to La Rochefoucauld's matching maxim, "Weakness, rather than virtue, is vice's adversary" (number 445), and then forward to Nietzsche's *The Will to Power*: ". . . the seduction that everything extreme exercises: we immoralists, we are the most extreme" (number 749).

The unconstrained tone of these maxims makes them highly enticing, particularly to young minds, as Pater understood. Yet such maxims are unlikely to pass Kant's fundamental test that a truly wise maxim is one that everyone should be able to follow. In these cases, if everyone did, the result would hardly look like a livable society.

extreme and disastrous cases, we reap not "ecstasy," but serial killers. Writings like the grisly novels of Sade and Bret Easton Ellis, along with splatter movies and TV programs, fulfill Pater's program of "getting as many pulsations as possible into the given time." Is there any reason why we should welcome them? For, contrary to my Jimmy Walker epigraph for Chapter VII ("No girl was ever seduced by a book"), books and images wield strong powers of seduction. One man's "hard gemlike flame" may light unpredictable fires in the neighborhood.

Less benighted authors, such as Flaubert and Tolstoy and Dostoyevsky, employ narrative perspective and complex characters to examine and criticize the ideal of pure experience. In *Crime and Punishment*, Sonia's quiet voice of Christian love enfolds Raskolnikov and finally dispels his aspirations to superhuman deeds. Her presence affects even Svidrigailov's concentrated evil. Literature provides many responses to the impulse toward experience. In a large number of cases, characters follow the downward path to wisdom that I describe at the end of Chapter II in discussing Milton. Therefore, they pass through the essential stage of experience on the way to wisdom. But it remains a stage, not the end in itself described by Pater.

Are there, then, things we should not know? Religion generally answers yes. Philosophy generally answers no. The history of curiosity as I trace it in earlier chapters and again in this chapter does not give so simple an answer. It holds out to us a profoundly cautionary tale about two new institutions that have risen up in competition with religion to accelerate our exploration of every domain of knowledge. Those institutions are science and art. Do they embody our most responsible behavior? Or organized presumption? History counsels us to learn patience in our quest for knowledge and to maintain "civilian control" over these two institutions. Dr. Faustus can all too easily metamorphose into Dr. Frankenstein.

In its innermost workings, literature carries an aching awareness of the veil of ignorance that accompanies our most intimate encounters with life. At unforeseen moments, consciousness may thwart our purposes. One index of vividness lies in the power of a work to create and sustain intervals of pregnant silence. Chekhov understood that the most intense moments onstage are wordless, breathless. Literature also carries an impulse toward experience for experience's sake, the yearning to experiment with life that drove Eve to eat the apple and Dr. Jekyll to create his disastrous alter ego, Mr. Hyde. The former led to the Fortunate Fall; the latter led to untold mayhem and self-destruction. The stories I have invoked encircle the terrain of forbidden knowledge without diminishing it. We cannot demonstrate finally that the Princesse de Clèves attained, or failed to attain, the goal of preserving her esteem for her husband and herself along with her love for the Duc de Nemours. But in some circumstances the path of abstinence may be fully responsible and life-affirming. The variety of such stories keeps before us the possibility that there are things we should not know by the flame of experience.

3. Last Tales

There is no end to the stories that reenact some aspect of forbidden knowledge. They may occur in the most abstract reaches of philosophical thought.

Philosophy loves to doubt itself. Some of the shrewdest thinkers destroy the foundations of their thought as fast as they lay them. "No man knows, or will ever know," declares the pre-Socratic Xenophanes, "the truth about the Gods and about everything I speak of." In the *Apology*, Socrates makes the unbeatable move of claiming that true wisdom lies in knowing the limits of wisdom. To state the predicament with a slightly different emphasis, if we have to justify our way of thinking before we start to think, we shall never start.* When Hegel, the most prolix of philosophers, reached this dilemma, he wrote tellingly about the dangers of examining the faculty of cognition, of turning thought back upon itself self-reflexively. Then, revealingly, Hegel concludes not with an argument but with a saying that is also a miniaturized story and an Irish bull. "To examine this so-called instrument [cognition, knowing]

*In a fine essay on Plato's metaphilosophy, Charles Griswold supplies the exact terms for the dilemma. "Metaphilosophy either leads us into an infinite regress or begs the question." Griswold also quotes the full passage from Hegel's *Logic* that I summarize in what follows.

is the same thing as to know it. But to seek to know before we know is as absurd as the wise resolution of Scholasticus not to venture into the water until he had learned to swim." Hegel's compact parable about learning to think or to swim or to do almost anything tells us both how foolish it can be to place limits on practical activities, and how necessary another kind of mental limit may be to prevent paralysis or damage when we try to reach the foundations of life or mind. Just do it; don't think too much about how to do it before you try. To make that simple point, Hegel the philosopher briefly becomes Hegel the narrator of an old wives' tale. Thus, tacitly and gracefully, he acknowledges the limits of philosophy and circles back to Meno's question to Socrates about how we can seek a thing about which we know nothing at all, not even a name.

Most stories of forbidden knowledge, however, are there waiting where we expect to find them. The second sister in a fairy tale chooses to marry a wealthy man with a terrifying blue-black beard and many fine houses. No one knows what became of his previous wives. After a month of marriage, he leaves for a long journey. Bluebeard entrusts to his wife the keys to his castle and forbids her to enter one small basement room. Visits from family and friends cannot distract her from her curiosity. When she finally opens the forbidden door, she finds the ghastly remains of the previous wives. Unable to clean the bloodstain from the key, she cannot hide her trespass from Bluebeard on his return. He says she must die. She asks for time to pray and sends her sister, Anne, to the tower to see if her brothers are coming as expected. They arrive just in time to save her and kill Bluebeard. She forgets him in a happy marriage to a good man.

Perrault's Mother Goose story of 1697 has several folk elements but remains to a large extent his own creation. One cannot miss the Wife of Bath effect. Bluebeard's evil nature is signaled both by his "ugly and terrible beard" and by his missing wives. The second sister makes the grave error of accepting him for his wealth and ignoring his character. She is saved from evil by equally evident goods: religion (even if she prays only to stall for time) and family. Those forces finally overcome Bluebeard. Forgiven, she is permitted another life, and justice is done.

At the end of his prose fable, Perrault appends two "morals" in verse. Curiosity always lands you in trouble. Sexual jealousy causes great harm. In his comments in *The Uses of Enchantment,* Bruno Bettelheim does not differ much from Perrault. "However one interprets 'Bluebeard,' it is a cautionary tale which warns: Women, don't give in to your sexual curiosity: men, don't permit yourself to be carried away by your anger at being sexually betrayed" (302). Susan Brownmiller would be impatient with this bland cover-up. For her, the tale has a hidden meaning, revealing its true source, Gilles de Rais.* This historical figure, Joan of Arc's stalwart first lieutenant, went on to abduct, rape, and murder scores of young boys. "Bluebeard" carries within it the ancient myth of the heroic rapist, Brownmiller maintains, and it celebrates that myth by veiling it in an innocent-looking children's story. But Brownmiller has read tendentiously.

To my mind, the tale of seemingly idle curiosity reveals rather than veils the motif of violent sexual crime. And the crimes are not condoned: they are punished. The young wife is saved from death and given another chance. "Bluebeard" is not a Trojan horse introducing dangerous forces into the polis and our psyche. Like healthy fairy tales, it attenuates the full virulence enough to warn against and perhaps to immunize against sexual terrorism and against sexual curiosity. There are some things we should not investigate.

The sinister element that lurks below the surface of "Bluebeard" lies right out in the open in the ancient myth of the Sphinx. This monster with the body of a griffin, the wings of a bird, and the face and voice of a virgin preys on travelers near Thebes. She poses riddles supplied to her by the Muses. Œdipus solves one of her riddles, overcomes her, and as a result goes on to fulfill his own tragic fate. The Sphinx resembles Bluebeard in that she is a predator associated with a mystery. The story of the Sphinx and Œdipus comes close to being a foundation story like Cain and Abel, or Romulus and Remus, for Œdipus saves the city of Thebes from a terrible scourge. His use of knowledge to crack the riddle serves a

---

*Without mentioning him, Brownmiller has picked up this claim from Jules Michelet's *Histoire de France* (1833–1867). Perrault scholars like Jacques Barchilon and Günther Lontzen remain skeptical about Michelet's statement, for he supplies no source.

beneficent purpose, but his hubris here, as in earlier episodes, condemns him to his own downfall and punishment.

My discussion in Chapter VI of Francis Bacon's version of the Sphinx story (see Appendix III) deals with its covert warning against the traditional distinction between pure and applied research. In three cryptic pages, the patron of modern inductive science expresses the proposition that science is a potentially dangerous monster that both occupies the lofty places of knowledge and "infests the roads" to challenge mortals with cruel questions. Sphinx-science poses two kinds of riddles: about the nature of things and about the nature of men. Bacon implies that questions about the nature of men—what he calls elsewhere "proud knowledge," almost a euphemism for forbidden knowledge—bring the real danger, like the question posed to Œdipus. The moral Bacon draws from this tale pulls us back to the motif of scale and pace, of *portée* and reach, of taking our time in approaching ultimate questions like the secrets of life and of mind. "Nor is that other point to be passed over, that the Sphinx was subdued by a lame man with club feet: for men generally proceed too fast and in too great a hurry to the solution of the Sphinx's riddles." Let us beware of crash programs and reductionist solutions. If the Sphinx represents science in its most dangerously alluring form, then we must find the courage to resist her riddling challenges, to tame her, rather than to be devoured by her.

Another fabulous creature lurks in our vicinity, an animal less monstrous than mysterious. The Unicorn has the body of a horse and bears on its forehead one long, straight, spiral multicolored horn, measured in cubits. Of Indian origin, the Unicorn became a figural element in Scottish heraldry and thus found its way back via colonization into the Indian coat of arms. One is justified in thinking of the Unicorn as the opposite of the scapegoat. Instead of carrying off our sins on its head into the wilderness, it brings in from the wilderness an elusive purity and is drawn to a gentle virgin. Its ludicrous horn represents sheer ornamental display, an awkward impediment to movement, a displaced male member, an antenna, a potential weapon, and a symbol of election and power.

But the Unicorn stands apart: It is a creature still without a story, without a full identity. It figures in thousands of images and narratives, but no one has yet discovered or invented its legend. The single horn refers less to battle than to magic and the erotic. Mythological and pseudoscientific acounts (in the Greek Ctesias and Aristotle, in the Old Testament creature called Re'em, in medieval bestiaries, and in Sir Thomas Browne) all portray the animal as appearing and disappearing in unpredictable ways, more hidden than sighted.* Allegedly, the Unicorn has been hunted and killed, captured and tamed. Sometimes depicted in combat, the Unicorn is more usually shown in scenes of worship, sacrifice, and domesticated love. But cumulatively, through the many fragmentary appearances that make up our knowledge of the shadowy beast, the Unicorn does not reveal a clear meaning, either beneficent or sinister. It represents an enigma that no knowledge or interpretation can decipher.

I believe that the Unicorn may come to represent the other new realm to which we assign many spiritual and redemptive powers formerly belonging to religion: the realm of art. But have we fully tamed this handsome beast with the awkward horn? Should we be ready now to follow the arts wherever they lead us in the name of freedom and experience, of imagination and transgression and mystery? I respond that we would do well to watch over the Unicorn of aesthetic experience as attentively as we watch over the Sphinx of science. Bereft of a complete fable, the Unicorn has earned a place in our imagination as an *arcanum*, an emblem of what we do *not* know. Might it represent a benign version of the predatory Sphinx? It is too soon to say. Every day, the arts enter new domains and new media. We cannot tell in what proportion the resulting works will enlighten, or entertain, or infect. Meanwhile, we have moved a long way from the disinterestedness that gave fresh impetus to art and to science in the seventeenth and eighteenth centuries. To curiosity have been added since then the strong entangling factors of progress, free enterprise, compulsive consumerism, and a semiautonomous technology.

Sphinx and Unicorn have approached very close—to us and to each other. From their incipient mating spring vertiginous modes of experience, running from virtual reality to designer genes to mutual assured destruction. "After such knowledge, what forgive-

*In *The Animal That Never Was,* Matti Megged offers a careful illustrated history of the Unicorn and cites the earlier scholarly literature.

ness?"—T. S. Eliot's aching question in "Gerontion"—prompts us to look hard at the contours of forbidden knowledge and forbidden experience, both ancient and emerging, in the shifting landscape we inhabit. The time has come to think as intently about limits as about liberation. We walk warily between Sphinx and Unicorn.

Appendix I: Six Categories of Forbidden Knowledge

■

My subject is too extensive and elusive to submit readily to system and theory. Nevertheless, a few categories have emerged in the course of my assembling these stories. Some sorting out of differences will allow me to take stock of earlier discussions and to seek the beginnings of order in so great a variety.

I propose six categories of forbidden knowledge.

- Inaccessible, unattainable knowledge
- Knowledge prohibited by divine, religious, moral, or secular authority
- Dangerous, destructive, or unwelcome knowledge
- Fragile, delicate knowledge
- Knowledge double-bound
- Ambiguous knowledge

The categories overlap one another and also leave discernible gaps. But this recapitulation offers a modified perspective on both familiar and unfamiliar materials. The first four categories should be reasonably clear. The last two deal with features that will be harder to distinguish.

INACCESSIBLE, UNATTAINABLE KNOWLEDGE

Some aspects of the cosmos—of "reality"—cannot be reached by human faculties. That inaccessibility springs either from the inadequacy of human powers or from the remoteness of realms presumed to exist in ways inconceivable to us. We do not have to choose between the two epigraphs for Chapter VIII. Both are true. But Socrates goes deeper than Einstein. Socrates' words prepare the way for Pascal's wager and Huxley's coinage of *agnostic.** Einstein's words draw a comic paradox out of Pascal's insistence that we know our reach, our *portée,* between the two infinities that escape us.† My third epigraph for this book—*Individuum est ineffabile*—restricts us even more severely by implying that we cannot know even the particulars that lie closest to us, including ourselves.

The Judaic tradition of never uttering any name for the divine being represents his ineffability and his unknowability as categorical assumptions. In the mid-fifteenth century, the Catholic theologian Nicholas of Cusa wrote an influential book, *De docta ignorantia,* or *On Learned Ignorance.* "Absolute Truth is Beyond our Grasp," declares the title of his third chapter. The only way we can apprehend God is through faith operating as a negative theology of wisely looking the other way—learned ignorance. Mystics such as Eckhart and St. John of the Cross profess a similar faith in the unattainable and the unutterable. Kant's *Ding an sich*—the noumenous Thing in Itself—may or may not exist. In any case phenomena or appearances that we *are* able to know in space and time will never lead us to noumena. A contemporary philosopher, Colin McGinn, leans partially on Kant's noumenalism in developing a position he calls "the insolubility thesis," or "cognitive pessimism." McGinn's book *Problems of Philosophy: The Limits of Inquiry* argues that human thinking is essentially unsuitable to grasp the

*On Huxley's coinage, see pages 37-40.

†See page 29.

existence and nature of states like consciousness and free will. He sidesteps the traditional question of knowledge by arguing that our access to truth may lie not in any faculty like reason, but in our genes. Untroubled by any Coppelia complex, McGinn welcomes the thought that we may all be automated.

Modern science contributes a number of illustrations of this category. In relativity theory, for example, one cannot refer to any universal *now* because any meaningful sense of simultaneity is limited by the finite velocity of light. We simply cannot know what is happening *now* on a distant star until its light signal reaches us after millions of (light-) years.

In this category it is simply the nature of things, including ourselves, that prevents us from knowing everything. Even Einstein's optimism concedes this final ignorance.

Knowledge Prohibited by Divine, Religious, Moral, or Secular Authority

Adam and Eve, Prometheus, and Psyche contravene a prohibition. These classic stories relate the consequences of powerful impatience struggling against even more powerful interdiction. Similar motifs recur in modified form in most quest stories including Dante's *Divine Comedy* (Peter Damian's warning in the *Paradiso*) and the tales of King Arthur and his knights (Perceval is too obedient). One of the most compact versions of this form of knowledge emerges from Hawthorne's short story "Ethan Brand." That intrepid figure sets out to seek the unpardonable sin; he discovers that he has already committed it by undertaking such a quest. It is in this category that the Wife of Bath effect comes into play. The second epigraph for this book points with a smile to the perverse human tendency to transform prohibition into temptation.

For reasons that scientists and officials would probably attribute to the sanctity of nature or of humanity, we currently prohibit research that would modify the germ line of human inheritance. In the realm of commerce and invention, we have developed copyright law and patent law. Both systems establish regulations around intellectual property to permit its being published and exploited while at the same time protecting ownership for a reasonable period. Such limitations paradoxically serve openness and the exchange of knowledge. In order to reinforce the dignity of the autonomous individual, privacy law, of comparatively recent origin, sets up restrictions on what others can learn about us and how far they can intrude upon us. The world is not a transparent medium of unrestricted observation and communication. The principle of privacy sets limits on what we can rightfully know about others' lives. But nothing remains secure for long. Information technology has already begun to infiltrate our privacy.

These ancient and modern prohibitions on particular areas of knowledge sometimes stimulate human curiosity more than they dampen it.

DANGEROUS, DESTRUCTIVE, OR UNWELCOME KNOWLEDGE

Playing with fire—or firearms—provides the most obvious and urgent example of dangerous knowledge. In Chapter VI, I consider the atomic bomb, recombinant DNA, and the Human Genome Project as representing this category of forbidden knowledge. We have learned to fear the effects that developing technology may have on the Earth's environment. In writing *Frankenstein,* still close to adolescent fantasy, Mary Shelley aimed not at the environmental but at the human depredations of scientific hubris. In comparison to her insistently cautionary tale, Goethe's *Faust* floats in ambivalence. Faust's appetite for sheer experience in the Gretchen episode and his technological experiments in draining swamps strew damage and suffering in his wake. Yet the Lord saves him at the end—for always striving. How shall we read this immense patchwork of a play? Faustian man properly has as many detractors as admirers in our day.

Unlike *Frankenstein,* there is nothing cautionary about the Marquis de Sade's writings. Rather than execrate, they embody the cruelty, sexual mayhem, and generalized killing that he preaches as a way of life for the rich and powerful. Those critics who find literary and moral virtues in Sade's work have much to answer for.

Simple prudence should impel us to take careful account of such forms of dangerous knowledge. Like drugs and tobacco, they need careful labeling and, if they reach commercial broadcast media, judicious regulation.

Fragile, Delicate Knowledge

The earlier chapter on *La Princesse de Clèves* and Emily Dickinson's veil poem examines forms of knowledge so sensitive that they may crumble and disappear in the moment of realization. One must approach one's own and others' deepest feelings and yearnings with circumspection for fear of driving them into hiding. The symbolist and decadent aesthetic at the close of the nineteenth century favored withdrawal from full-fledged experience and took refuge in a refined realm of language and imagination. In the poem "Art poétique" Verlaine chooses musicality, nuance, and veiled beauty out of which to compose his *chanson grise.*

For certain men and women, the sexual response falls into a delicate area far removed from conquest and aggressiveness. Some highly responsive men, for whom rape is unthinkable, reach full sexual arousal in circumstances that never exclude the possibility of fiasco. Not violence but tenderness serves their appetites.

A comparable discrimination of effects has long existed in writing published under the threat of persecution for heretical views. Before the seventeenth century in Europe, Leo Strauss observes, many original thinkers "wrote between the lines" in order to allow alert readers to "catch a glimpse of forbidden fruit." The essential teaching often lay concealed inside a protective garment, and one gained access to it by patient interpretation. Maimonides' *Guide for the Perplexed* approached prohibited subject matter—the secrets of the Hebrew Bible—by employing the hints and indirections of esoteric writing. In some circumstances, the truth survives better veiled than naked. At the lowest order of magnitude in physics, particles or waves become so sensitive that the act of observation affects their energy level and modifies the reading. We do not know in advance whether our approach to something or someone will destabilize or even desecrate the hoped-for response. Any TV camera crew can observe how its mere presence on the scene modifies the nature of the events it was sent to record.

Fragile knowledge finds its natural home in the domains of discretion and privacy.

KNOWLEDGE DOUBLE-BOUND

The fifth category differs considerably from the others and will be harder to define. Both common sense and the history of philosophy recognize two kinds, two tendencies of knowledge. We may approach, enter into, sympathize with, and unite with the thing known in order to attain subjective knowledge. Or we may stand outside, observe, anatomize, analyze, and ponder the thing known in order to attain objective knowledge. Subjective or empathetic knowledge causes us to lose a judicious perspective on the object; objective knowledge, in seeking to maintain that perspective, loses the bond of sympathy. We cannot know something by both means at the same time. The attempt to reconcile the two or to alternate between them leads to great mental stress. Orestes recoiled from his objective duty to avenge his father, Agamemnon, because of his subjective revulsion to killing his mother, Clytemnestra. In explaining how best to comprehend the sublime magnitude of the Great Pyramids in Egypt, Kant wrote with startling simplicity. "We must avoid coming too near just as much as remaining too far away" (*Critique of Judgment*, I, 26). Flaubert was less judicious. "The less one feels a thing the more apt one is to express it as it is" (letter to Louise Colet, March 4, 1852).

For the Romantics in their reaction to Enlightenment reason, the distinction between the two modes of knowledge appeared to reach even deeper within us. Schiller devoted his sixth letter, *On the Aesthetic Education of Man*, to the dissociation of reason from feeling or imagination. "It was civilization itself which inflicted this wound upon modern man." Wordsworth discovered a similar division in the mind.

> *The groundwork, therefore, of all true philosophy is the full apprehension of the difference between . . . that intuition of things which*

> *arises when we possess ourselves, as one with the whole . . . and that which presents itself when . . . we think of ourselves as separated beings, and place nature in antithesis to the mind. As object to subject.*
>
> (*The Friend*)

Wordsworth had picked up the terms from his friend Coleridge. Thomas Carlyle delighted in mocking Coleridge's constant return in his rambling monologues to the snuffled words *om-m-mject* and *sum-m-mject.* A sentence in Chapter XII of Coleridge's intellectual autobiography, *Biographia Literaria,* makes thinking sound easy. "During the act of knowledge itself, the objective and subjective are so instantly united, that we cannot determine to which of the two priority belongs." The remainder of the chapter removes that impression of ease.

A powerful discussion of this double-bind blocking us from balanced or whole knowledge overwhelms the concluding chapters of Lévi-Strauss' anthropological narrative *Tristes Tropiques* (1955). Having devoted five years to fieldwork among isolated Indians in Brazil, he finds himself the victim of a "mental disorder." He has become lost, suspended between two cultures. Insofar as he has entered, as ethnographer, into the Indian culture he is studying, he has lost track of his own culture and of the hard-earned scientific disciplines that led him to this enterprise. Insofar as he remains detached from the culture under study, he lacks essential connections with it that would permit full understanding. Lévi-Strauss' final pages develop a crescendo of tragic meditation over his double bind. "There is no way out of the dilemma." His "sin" is to be bound to two cultures, and, therefore, to none. He calls his predicament an "abyss," out of which he can communicate with no one—except perhaps a cat. The "tropics" of his book's title are profoundly "sad" for Lévi-Strauss because they represent this personal and professional double bind. He paints it as lurking behind every inquiry of the mind into the nature of the world and the people in it. This form of forbidden knowledge known to the boldest explorers and the subtlest investigators implies a deep incompatibility between the human mind and the world around it—the converse of Einstein's happier observation: "The most incomprehensible thing about the universe is that it is comprehensible."

The character Kurtz in Conrad's *Heart of Darkness*, probing the limits of savagery in Africa, flings himself across this abyss of unknowing and sacrifices his humanity. It is the response of desperation.

We can discern an even more sustained effort than Lévi-Strauss' to surmount the conflict between objective and subjective knowledge in William James' *The Varieties of Religious Experience* (1902). By using a case-history approach similar to the narrative style of *Tristes Tropiques,* James moves as close as he can to the religious experiences that concern him. He succeeds in showing a profound sympathy toward alien feelings without renouncing his detachment. But in the final chapter, he acknowledges a frustration similar to Lévi-Strauss', though with less hyperbole. Referring to religion and mysticism, James concludes laconically that "Knowledge about a thing is not the thing itself."

Eight lines further on in the same passage, James suddenly and without explanation quotes in French the proverb "To understand is to forgive."* What can he possibly have in mind? I believe James is here calling our attention fleetingly to the other side of the double bind. Exterior objective knowledge will never carry us to a full grasp of any subjective experience. On the other hand, as the French proverb suggests, full empathy with another experience or another life takes away from us the capacity to see it objectively and to judge it aright. My discussion of *Billy Budd* and *The Stranger* in Chapter V deals at some length with this interference in the reader's mind between one form of knowledge and the other. Each novel carries us so close to the principal character that we run the risk of being unable to form a judicious evaluation of the homicide he perpetrates. This fifth form of forbidden knowledge arises from a familiar fissure at the heart of our thinking. Hard as we may try, we cannot be both inside and outside an experience or a life—even our own.

---

*James gives a common variation: *Tout savoir c'est tout pardonner.*"

Ambiguous Knowledge

I have not finished with the paradoxes that affect knowing, for it is necessary to follow where the stories lead. By "ambiguous," I refer to a condition in which what we know reverses itself right under our noses, confounds us by turning into its opposite.

Take the end of *Paradise Lost.* Adam and Eve have repented of their sin and been granted "many days" of mortal life—but not in Paradise. Then the Archangel Michael leads Adam to a hilltop and shows him the future, including the coming of Christ and his redemption of Adam's sin. Adam feels both "joy and wonder" over a change he cannot understand.

> *"O goodness infinite, goodness immense!*
> *That all this good of evil shall produce,*
> *And evil turn to good . . ."*
>
> (XII, 469–71)

Adam's universe has been totally transformed. Milton's lines represent the best-known literary expression of the Fortunate Fall, a contradiction or reversal of interpretation that had been gradually adopted as Christian doctrine during the Middle Ages.*

With no reference to Milton, the philosopher's philosopher Kant concocted his own secular version of the reversal. In "Conjectural Origins of the Human Story" (1786), he writes as if he were being interviewed as the author of a playful novel called *Adam and Eve.* Kant explains that reason and imagination, the secular virtues of his "flight of fancy," finally bring about a "fall" as double-edged as Milton's. "For the individual, who in the use of his freedom has regard only for himself, such a change was a loss; for nature, whose end for man concerns the species, it was a victory." In this short essay, Kant becomes sly and lighthearted enough to recast the Adam and Eve story.

Such a reversal of effect turns up in other places: the principle

*No account surpasses that of A. O. Lovejoy's succinct 1939 essay "Milton and the Paradox of the Fortunate Fall."

of vaccination; the Wife of Bath effect; the Eldorado reaction.* In these cases, respectively, poison or infection turns into remedy; what is forbidden becomes desirable; the ideal becomes intolerable. We come up against a pun or ambiguity in the very nature of things.

These forms of double meaning leave us confounded by paradox. Our mind reckons uncomfortably with contradiction affirmed. The fact that such a contradiction lies at the heart of Christian doctrine, of our immune system, and of other crucial human activities opens an area of uneasy knowledge. Under rare circumstances, *A* is not exclusively *A*: *A* is *B* while remaining *A*. We enter this chameleon world warily. The Wife of Bath's "Forbede us thyng, and that desiren we" reports on the unstable human condition that John Locke looked out on from the other side: "Where there is no law there is no freedom."

Two further instances of ambiguous knowledge insist on being heard. Do writers fare best under repression and persecution or in a free society? After Eastern European countries regained independence and the Soviet Union came to an end about 1990, respect for dissident literature diminished rapidly, and writers found their role difficult to reestablish in a market economy. At a 1992 *Partisan Review* conference on intellectuals in Eastern Europe, Saul Bellow cut through to the essential dilemma by wondering "whether we need these colonial evils of dictatorship to keep us honest." Years earlier, the Cuban dissident Herberto Padilla had turned the paradoxical situation into an incipient proverb. "The best poems have always been born beneath the jailer's lamp." We shall not soon learn what combination or alternation of freedom and repression will make writers honest and responsible.

The second instance of ambiguous knowledge concerns a double duty that affects each one of us. We need to be faithful to our traditions and our knowledge, to our community and our history. And we also need to be able to respond with guarded flexibility and understanding to challenges to those traditions and that knowledge. To discharge that double duty without fanaticism while

---

*When he reaches the utopian country of Eldorado, Candide cannot abide the absence of outward conflict and the tranquillity of mind that characterize that sheltered land. In a similar and more complex response, Gulliver loses his mind on his fourth voyage to the purely reasonable society of the Houyhnhnms.

firmly maintaining a set of scruples based on reason and experience forms the challenge of an entire lifetime. How can we be faithful and unfaithful at the same time? Over and over again in the tiny decisions of everyday life, we must do just that at every level of action and reflection, through every fluctuation of doubt and faith.

Appendix II: The Occult

■

To many people, the term *forbidden* knowledge suggests, first of all, an area known as the occult. The word *occult* has long been used to designate a fluid collection of traditions and writings bordering closely on religion, magic, and superstition. Beyond its root meaning of secret or recondite, occult has a number of strong associations. It refers to secret truths of great antiquity, not of recent discovery. All its manifestations point to the existence of an ineffable spiritual being revealing itself through light and often through love. In the face of many separate churches and religions, these occult beliefs aspire to be universal and perennial. The most basic doctrines are of great simplicity and appeal. First, the universe has two parts: the material world of appearances and a higher spiritual truth hidden behind appearances. Second, the two parts are related through analogy (correspondences, symbols, affinities) revealed by heightened vision, magic, and prophecy. If you open your eyes properly, you may see that everything connects. That is the hidden knowledge. As a young philosopher of Romanticism, Schelling wrote an almost mathematical statement of the occultist point of view.

> *The analogy of each part of the universe to the whole is such that the same idea is reflected constantly from whole to part and part to whole. The analogies of the different parts of physical nature among themselves serve to establish the supreme law of creation, variety in unity, and unity in variety. What is more astonishing, for example, than the relation of sounds and forms, of sounds and colors?*

Schelling's words were quoted by Mme de Staël in *De l'Allemagne* (1810) and by Fourier in *Nouveau monde industriel et sociétaire* (1829) and left their mark on Nerval and Baudelaire and Emerson, among many others. I venture that few of us are insensitive to the beckoning of the occult in this loose form.

A powerful line of ancient figures has developed and transmitted the occult tradition: Hermes Trismegistus, a legendary Egyptian god figure projected backward into the origins of culture by third-century neo-Platonists; Zoroaster and Pythagoras in the seventh century B.C.; Simon Magus, a contemporary of Jesus and the first heretic, and Apollonius of Tyana in the first century C.E.; Agrippa, Paracelsus, and Nostradamus in the sixteenth century; and Cagliostro in the eighteenth. Their multifarious teachings intermingle with the most important single strand of occultism: the Hebrew Kabbalah, originating in the second century and revived in the thirteenth.* The Kabbalah taught that through an elaborate letter and number symbolism, Scripture can reveal to us the innermost secrets of the universe.

Within this tradition of mystical study under the Kabbalah, it is important to distinguish two paths. The exoteric or moderate path is open to all who devote themselves to serious Torah study. The esoteric or intensive path, on the other hand, is reserved for an elite few prepared to employ magic incantations and face unforeseen dangers. The analogy here is that of entering a sacred domain forbidden to the unprepared. Along this path of the Kabbalah, some initiates will reach a dazzling revelation of the divine in Ezekiel's vision of a throne-chariot attended by four winged creatures and

*See the writings of Gershom Scholem and Moshe Idel.

rolling on four wheels within wheels. Comparable status is reserved for the dozen cryptic lines from the Talmud about four men who entered the King's orchard. When they reached the stone fountain, one gazed and died, one gazed and lost his mind, one "cut down the shoots" (that is, fell into heresy), and one departed in peace. The Talmud deals also with safeguards on initiation into these abstruse and dangerous matters.

> *The Laws of incest may not be expounded to three persons, nor the Story of Creation before two persons, nor the subject of the Chariot before one person alone unless he be a Sage and comprehends of his own knowledge. Whoever puts his mind to these four matters it were better for him if he had not come into the World.*

Some subjects may lead to undesirable thoughts, and all the above matters may "cause a falling away from the true moral teaching." Maimonides' *Book of Knowledge*, widely studied among Muslims, Jews, and Christians, conveys the same message of carefully restricted access.

> *The ancient sages enjoined us to discuss these subjects only privately, with one individual, and then only if he be wise and capable of independent reasoning. In this case, the heads of the topics are communicated to him, and he is instructed in a minute portion of the subject. These topics are exceedingly profound; and not every intellect is able to approach them.*
>
> ("On Secrets," 36b–39b)

Now it is not hard to see how close to this occult tradition we must locate a number of influential modern figures. Swedenborg's world of spirits and correspondences and analogies revives the hidden doctrine of the Kabbalah. He passed it on not only to the Church of the New Jerusalem but also to such writers as Blake and Emerson and a whole generation of Romantic artists. After his fashion, Faust belongs to this legacy. Bored by a life of scholarship, he turns to magic and occultist formulae to liberate himself from dusty books and to attain direct experience. Romantic poets in all European languages sought forms of occult knowledge to further their engagement with spiritual powers. Yeats developed a system of spiritual beings in *A Vision*. Mallarmé spoke for these modern poets in his many defenses of obscurity in literature. He wrote that we need "systematic ways to protect the entrance to the temple . . . to ward off anyone who does not have enough love." Kandinsky's *Concerning the Spiritual in Art* (1912), which can be read as an occultist manifesto, has probably been the single most influential written work on the visual arts during the twentieth century.

The occult in its common acceptance is not a special category under forbidden knowledge, but a vast catchall collection of religious, secular, psychic, and magical lore, mostly ancient. A popular and unscholarly compendium is called *Zolar's Encyclopedia of Ancient and Forbidden Knowledge*. After a superficial chapter on the Kabbalah, it covers the astral world, the mysteries of sex, mind power, astrology (at great length), methods of winning, and much more. I have not dealt with these areas. Their association in popular thought has given to the occult so broad a meaning that it will neither correspond to nor contrast with forbidden knowledge as I am treating it.

APPENDIX III: "THE SPHINX" BY FRANCIS BACON

■

Sphinx, says the story, was a monster combining many shapes in one. She had the face and voice of a virgin, the wings of a bird, the claws of a griffin. She dwelt on the ridge of a mountain near Thebes and infested the roads, lying in ambush for travellers, whom she would suddenly attack and lay hold of; and when she had mastered them, she propounded to them certain dark and perplexing riddles, which she was thought to have obtained from the Muses. And if the wretched captives could not at once solve and interpret the same, as they stood hesitating and confused she cruelly tore them to pieces. Time bringing no abatement of the calamity, the Thebans offered to any man who should expound the Sphinx's riddles (for this was the only way to subdue her) the sovereignty of Thebes as his reward. The greatness of the prize induced Œdipus, a man of wisdom and penetration, but lame from wounds in his feet, to accept the condition and make the trial: who presenting himself full of confidence and alacrity before the Sphinx, and being asked what kind of animal it was which was born four-footed, afterwards became two-footed, then three-footed, and at last four-footed again, answered readily that it was man; who at his birth and during his infancy sprawls on all fours, hardly attempting to creep; in a little while walks upright on two feet; in later years leans on a walking-stick and so goes as it were on three; and at last in extreme age and decrepitude, his sinews all failing, sinks into a quadruped again, and keeps his bed. This was the right answer and gave him the victory; whereupon he slew the Sphinx; whose body was put on the back of an ass and carried about in triumph; while himself was made according to compact King of Thebes.

The fable is an elegant and a wise one, invented apparently in allusion to Science; especially in its application to practical life. Science, being the wonder of the ignorant and unskilful, may be not absurdly called a monster. In figure and aspect it is represented as many-shaped, in allusion to the immense variety of matter with which it deals. It is said to have the face and voice of a woman, in respect of its beauty and facility of utterance. Wings are added because the sciences and the discoveries of science spread and fly abroad in an instant; the communication of knowledge being like that of one candle with another, which lights up at once. Claws, sharp and hooked, are ascribed to it with great elegance, because the axioms and arguments of science penetrate and hold fast the mind, so that it has no means of evasion or escape; a point which the sacred philosopher also noted: *The words of the wise are as goads, and as nails driven deep in.* Again, all knowledge may be regarded as having its station on the heights of mountains; for it is deservedly esteemed a thing sublime and lofty, which looks down upon ignorance as from an eminence, and has moreover a spacious prospect on every side, such as we find on hill-tops. It is described as infesting the roads, because at every turn in the journey or pilgrimage of human life, matter and occasion for study assails and encounters us. Again Sphinx proposes to men a variety of hard questions and riddles which she received from the Muses. In these, while they remain with the Muses, there is probably no cruelty; for so long as the object of meditation and inquiry is merely to know, the understanding is not oppressed or straitened by it, but is free to wander and expatiate, and finds in the very uncertainty of conclusion and variety of choice a certain pleasure and delight; but when they pass from the Muses to Sphinx, that is from contemplation to practice, whereby there is necessity for present action, choice, and decision, then they begin to be painful and cruel: and unless they be solved and disposed of they strangely torment and worry the mind, pulling it first this way and then that, and fairly tearing it to pieces. Moreover the riddles of the Sphinx have always a twofold condition attached to them: distraction and laceration of mind, if you fail to solve them; if you succeed, a kingdom. For he who understands his subject is master of his end; and every workman is king over his work.

Now of the Sphinx's riddles there are in all two kinds: one concerning the nature of things, another concerning the nature of man; and in like manner there are two kinds of kingdom offered as the reward of solving them; one over nature, and the other over man. For the command over things natural,—over bodies, medicines, mechanical powers, and infinite other of the kind—is the one proper and ultimate end of true natural philosophy; however the philosophy of the School, content with what it finds, and swelling with talk, may neglect or spurn the search after realities and works. But the riddle proposed to Œdipus, by the solution of which he became King of Thebes, related to the nature of man; for whoever has a thorough insight into the nature of man may shape his fortune almost as he will, and is born for empire; as was well declared concerning the arts of the Romans,—

Be thine the art,
O Rome, with government to rule the nations,
And to know whom to spare and whom to abate,
And settle the condition of the world.

And therefore it fell out happily that Augustus Cæsar whether on purpose or by chance, used a Sphinx for his seal. For he certainly excelled in the art of politics if ever a man did; and succeeded in the course of his life in solving most happily a great many new riddles concerning the nature of man, which if he had not dexterously and readily answered he would many times have been in imminent danger of destruction. The fable adds very prettily that when the Sphinx was subdued, her body was laid on the back of an ass: for there is nothing so subtle and abstruse, but when it is once thoroughly understood and published to the world, even a dull wit can carry it. Nor is that other point to be passed over, that the Sphinx was subdued by a lame man with club feet; for men generally proceed too fast and in too great a hurry to the solution of the Sphinx's riddles; whence it follows that the Sphinx has the better of them, and instead of obtaining the sovereignty by works and effects, they only distract and worry their minds with disputations.

(from *The Wisdom of the Ancients*, 1610)

Acknowledgments

■

My warm thanks go to friends and colleagues at Boston University and elsewhere who have helped me by suggesting materials, discussing knotty problems, and criticizing portions of the manuscript: Harold B. Alexander, Horace Allen, Rémi Brague, Frederick Brown, Jane Brown, Donald Carne-Ross, Kathe Darr, Joan Daves, Lewis Feuer, Abigail Gillman, Elizabeth Goldsmith, Wolfgang Haase, Kenneth Haynes, Geoffrey Hill, Evelyn Fox Keller, William Kerrigan, Roger Kimball, Joc Kirchberger, Jeffrey Mehlman, James Miller, Michael Prince, Christopher Ricks, Frederick Schauer, James Schmidt, Abner Shimony, Alfred Tauber, Rosanna Warren, Hellmut Wohl, and—through thick and thin—William Wise.

I am also grateful to colleges and universities whose invitations to lecture gave me opportunities to try out sections of this book: Agnes Scott College, Baldwin-Wallace College, Boston University, Brown University, Indiana University (Patten Lecture Series), University of Iowa, University of Oregon (Humanities Center), University of New Hampshire, and University of Vermont.

Bibliography

■

Standard and classical works existing in many editions have not been listed. These include works by Homer, Plato, Aristotle, Aquinas, Dante, Francis Bacon, Descartes, Milton, Mme de Staël, and George Eliot.

Abrams, M. H. *Natural Supernaturalism: Tradition and Revolution in Romantic Literature.* New York: W. W. Norton, 1971.

L'affaire Sade. Compte-rendu du procès intenté par le Ministère Public. Paris: Pauvert, 1957.

Aloff, Mindy, "The Company He Kept" *The New Republic,* August 1, 1994.

Alter, Robert and Frank Kermode. *The Literary Guide to the Bible.* Cambridge: Harvard University Press, 1987.

Apollinaire, Guillaume. "The Divine Marquis." Introduction to *L'œuvre du Marquis de Sade.* Paris: Bibliothéque des Curieux, 1909.

Arendt, Hannah. *Eichmann in Jerusalem: A Report on the Banality of Evil.* Rev. ed. New York: Viking Press, 1964.

Artz, Frederik. *From the Renaissance to Romanticism: Trends in Style in Art, Literature, and Music 1300–1830.* Chicago: Chicago University Press, 1962.

Asimov, Isaac, et al., eds. *Machines That Think.* New York: Holt, Rinehart, and Winston, 1983.

Attorney General's Commission on Pornography. *Final Report.* Washington, D.C.: U.S. Department of Justice, July 1986.

Barlow, Nora Darwin, ed. *The Autobiography of Charles Darwin, 1809–1882.* Bollingen Series. New York: Pantheon Books, 1958.

Barnstone, Willis, ed. *The Other Bible.* San Francisco: Harper and Row, 1984.

Barthes, Roland. *Sade, Fourier, Loyola.* Paris: Seuil, 1971.

Bataille, Georges. *Histoire de l'érotisme.* Paris, 1951.

———. *La littérature et le mal.* Paris: Gallimard, 1976.

———. *L'érotisme.* In *Œuvres complètes.* Vol. 10. Paris: Gallimard, 1987.

———. *Visions of Excess: Selected Writings 1927–1939.* Edited by Allan Stoekl. Minneapolis: University of Minnesota Press, 1985.

Beauvoir, Simone de. *Privilèges.* Paris: Gallimard, 1955.

Beckett, Samuel. *Proust.* New York: Grove Press, n.d.

Bell, Millicent, "The Fallacy of the Fall in *Paradise Lost.*" *PMLA.* (September 1953).

Benedict, Ruth. *Patterns of Culture.* Boston: Houghton Mifflin, 1934.

Bénichou, Paul. *Le sacre de l'écrivain, 1750–1830.* Paris: Corti, 1993.

Berlin, Isaiah. *The Crooked Timber of Humanity: Chapters in the History of Ideas.* Edited by Henry Hardy. New York: Alfred A. Knopf, 1991.

———. "Historical Inevitability" and "Two Concepts of Liberty" in *Four Essays on Liberty.* New York: Oxford University Press, 1969.

———. Introduction to *The Age of Enlightenment: The Eighteenth-Century Philosophers.* Boston: Houghton Mifflin, 1956.

Berman, Marshall. *All That Is Solid Melts into Air: The Experience of Modernity.* New York: Simon and Schuster, 1982.

Bernstein, Jeremy. *Quantum Profiles.* Princeton: Princeton University Press, 1991.

Bettelheim, Bruno. *The Uses of Enchantment: The Meaning and Importance of Fairy Tales.* New York: Alfred A. Knopf, 1976.

Bibesco, Martha. *Au bal avec Marcel Proust.* Paris: Gallimard, 1928.

Bishop, Jerry E., and Michael Waldholz. *Genome: The Story of the Most Astonishing Scientific Adventure. . . .* New York: Simon and Schuster, 1990.

Blumenberg, Hans. *The Legitimacy of the Modern Age.* Translated by Robert M. Wallace. Cambridge: MIT Press, 1983.

Boerner, Peter, and Sidney Johnson, eds. *Faust through Four Centuries: Retrospect and Analysis.* Tübingen: Niemeyer Verlag, 1989.

Bronowski, Jacob. *Science and Human Values.* New York: Harper and Brothers, 1956.

Brown, Jane K. *Goethe's Faust: The German Tragedy.* Ithaca, N.Y.: Cornell University Press, 1985.

Brownmiller, Susan. *Against Our Will: Men, Women, and Rape.* New York: Simon and Schuster, 1975.

Bugliosi, Vincent. *Helter Skelter: The True Story of the Manson Murders.* New York: W. W. Norton, 1974.

Bury, J. B. *The Idea of Progress: An Inquiry into Its Origin and Growth.* New York: Dover Publications, 1955.

Butler, E. M. *The Myth of the Magus.* Cambridge: Cambridge University Press; New York: Macmillan, 1948.

Camus, Albert. *The Stranger.* Translated from the French by Matthew Ward. New York: Alfred A. Knopf, 1988.

———. "Melville." In *Les écrivains célèbres.* Paris, 1953.

———. *The Rebel.* New York: Vintage Books, 1956.

Carlyle, Thomas, "Coleridge." In *The Life of John Sterling.* London: Chapman and Hall, 1851.

Carter, Angela. *The Sadeian Woman.* New York: Pantheon Books, 1978.

Cassirer, Ernst. *Rousseau, Kant and Goethe.* Princeton: Princeton University Press, 1945.

———. *An Essay on Man: An Introduction to a Philosophy of Human Culture.* New Haven: Yale University Press, 1944.

Chambers, Frank, P. *The History of Taste: An Account of the Revolutions of Art Criticism and Theory in Europe.* New York: Columbia University Press, 1932.

Changeux, Jean-Pierre. *Neuronal Man: The Biology of Mind.* Translated by Laurence Garey. New York: Pantheon Books, 1985.

Châtelet, Noëlle, ed. *Sade: Système de l'agression, textes politiques et philosophiques.* Paris: Aubier-Montaigne, 1972.

Chesneaux, Jean. *The Political and Social Ideas of Jules Verne.* Translated from the French by Thomas Wikeley. London: Thames and Hudson, 1972.

Clark, Kenneth. *The Nude: A Study in Ideal Form*. Bollingen Series. New York: Pantheon Books, 1956.
Collège de Sociologie (1937–39). Présenté par Denis Hollier. Paris: Gallimard, 1979.
Collingwood, R. J. *The Idea of History*. London: Oxford University Press, 1946.
Cover, Robert M. *Justice Accused: Antislavery and the Judicial Process*. New Haven: Yale University Press, 1975.
Cranston, Alan. "The Non-Event." *The New Republic*, August 21, 1995.
Crocker, Lester G. *Nature and Culture: Ethical Thought in the French Enlightenment*. Baltimore: Johns Hopkins University Press, 1963.
Cryle, Peter. *Geometry in the Boudoir: Configurations of French Erotic Narrative*. Ithaca, N.Y.: Cornell University Press, 1994.
Curtius, Ernst Robert. *European Literature and the Middle Ages*. Translated from the German by Willard R. Trask. Bollingen Series. New York: Pantheon Books, 1953.
Darnton, Robert. *The Forbidden Bestsellers of Pre-Revolutionary France*. New York: W. W. Norton, 1995.
Darwin, Charles. *The Expression of Emotions in Man and Animals*. 1872. Reprint, London: Julian Friedmann; New York: St. Martin's Press, 1979.
Davies, Sir John. *Nosce Teipsum*. 1599. See Sneath. *Philosophy in Poetry*.
Davis, Joel. *Mapping the Code: The Human Genome Project and the Choices of Medical Science*. New York: Wiley, 1990.
Della Mirandola, Pico. *On the Dignity of Man*. 1488.
Degler, Carl, N. *In Search of Human Nature: Decline and Revival of Darwinism in American Social Thought*. New York: Oxford University Press, 1991.
Dickinson, Emily. *The Complete Poems of Emily Dickinson*. Edited by Thomas H. Johnson. Boston: Little, Brown, 1960.
———. *The Letters of Emily Dickinson*. Edited by Thomas H. Johnson. 2 vols. Cambridge: Harvard University Press, 1958.
Dobzhansky, Theodosius. *Mankind Evolving: The Evolution of the Human Species*. New Haven: Yale University Press, 1962.
Dodds, E. R.. *The Greeks and the Irrational*. Berkeley and Los Angeles: University of California Press, 1951.
Donnerstein, Edward, Daniel Leinz, and Steven Penrod. *The Question of Pornography: Research Findings and Policy Implications*. New York: Free Press, 1987.
Donoghue, Denis. *Thieves of Fire*. New York: Oxford University Press, 1974.
Dostoevsky, F. M. *The Diary of a Writer*. Salt Lake City: Peregrine Smith Books, 1985.
DuBartas. *La Semaine*. 1578.
DuBois-Reymond, Emil. *Reden*. 2 vols. Leipzig, 1912.
Duster, Troy. *Backdoor to Eugenics*. New York: Routledge, 1990.
Dyson, Freeman, J. *Infinite in all Directions*. New York: Harper and Row, 1988.
Edelman, Gerald, M. *Bright Air, Brilliant Fire: On the Matter of the Mind*. New York: BasicBooks, 1992.
Eichner, Hans. "The Eternal Feminine: An Aspect of Goethe's Ethics." Reprinted in *Faust* by Johann Wolfgang von Goethe. Translated by Walter Arndt. Norton Critical Edition. New York: W. W. Norton, 1976.
Elias, Norbert. *The Civilizing Process: The History of Manners*. Translated by Edmund Jephcott. New York: Urizen Books, 1978.

Elster, Jon. *Ulysses and the Sirens: Studies in Rationality and Irrationality*. New York: Cambridge University Press, 1979.
Empson, William. *Milton's God*. London: Chatto and Windus, 1961.
Enright, D. J. *A Man for Sentences*. Boston: Godine Publishers, 1985.
Evans, J. M. *Paradise Lost and the Genesis of Tradition*. Oxford: Clarendon Press, 1968.
Forsyth, Neil. *The Old Enemy: Satan and the Comfort Myth*. Princeton: Princeton University Press, 1987.
Foucault, Michel. *Madness and Civilization: A History of Insanity in the Age of Reason*. New York: Random House, 1965.
———. *The Order of Things: An Archaeology of the Human Sciences*. New York: Random House, 1970.
Frazer, Sir James George. *The Golden Bough: A Study in Magic and Religion*. Abridged edition. New York: Macmillan, 1922.
Freud, Sigmund. *Totem and Taboo*. 1913.
———. "The Uncanny." 1919.
Gadamer, Hans-Georg. *Truth and Method*. Translated by Joel Weinsheimer and Donald G. Marshall. 2d ed. New York: Crossroad, 1992.
Gallop, Jane. *Intersections: A Reading of Sade with Bataille, Blanchot, and Klossowski*. Lincoln: Nebraska University Press, 1981.
Gardner, Martin, ed. *Great Essays in Science*. New York: Pocket Books, 1957.
Genette, Gerard. *Figures II*. Paris: Seuil, 1969.
Ginger, Ray. *Six Days or Forever? Tennessee v. John Thomas Scopes*. Boston: Beacon Press, 1958.
Ginzburg, Carlo. *Clues, Myths, and the Historical Method*. Translated by John Tedeschi and Anne Tedeschi. Baltimore: Johns Hopkins University Press, 1989.
Girard, René, "Camus's Stranger Retried." In *To Double Business Bound*. Baltimore: Johns Hopkins University Press, 1978.
Gombrich, E. H. *In Search of Cultural History*. Oxford: Clarendon Presss, 1969.
Goodchild, Peter. *J. Robert Oppenheimer: Shatterer of Worlds*. Boston: Houghton Mifflin, 1981.
Gorer, Geoffrey. *The Life and Ideas of the Marquis de Sade*. 3rd ed. Reprint, London: Greenwood, 1978.
Gould, Stephen Jay, "The Monster's Human Nature." *Natural History* (July 1994).
Griswold, Charles, "Plato's Metaphilosophy: Why Plato Wrote Dialogues." In *Platonic Readings/Platonic Writings*. Edited by Charles Griswold. New York: Routledge, 1988.
Guern, Darko. *Metamorphoses of Science Fiction*. New Haven: Yale University Press, 1979.
Gulik, R. H. van. *Sexual Life in Ancient China: A Preliminary Survey of Chinese Sex and Society from ca. 1500 B.C. till 1644 A.D.* Leiden: E. J. Brill, 1974.
Haeckel, Ernst. *The Riddle of the Universe: At the Close of the Nineteenth Century*. Translated by Joseph McCabe. New York: Harper and Brothers, 1901.
Hammer, Carl. *Goethe and Rousseau*. Lexington: University Press of Kentucky, 1973.
Hanser, Richard. *A Noble Treason: The Revolt of the Munich Students Against Hitler*. New York: Putnam, 1979.
Hawking, Stephen. *A Brief History of Time: From the Big Bang to Black Holes*. New York: Bantam, 1988.

Hayman, Ronald. *De Sade: A Critical Biography*. London: Constable, 1978.
Henry, Patrick, ed. *An Inimitable Example: The Case for the Princesse de Clèves*. Washington, D.C.: Catholic University of America Press, 1992
Hillel, Marc, and Clarissa Henry. *Of Pure Blood*. Translated by Eric Mossbacher. New York: McGraw Hill, 1976.
Holbrook, David, ed. *The Case Against Pornography*. La Salle, Ill.: Library Press, 1973.
Holton, Gerald. *Science and Anti-Science*. Cambridge: Harvard University Press, 1993.
Holton, Gerald and Robert S. Morison, eds. "Limits of Scientific Inquiry." *Daedalus*. (spring 1978).
Horkheimer, Max, and Theodore Adorno. *The Dialectic of Enlightenment*. Translated by John Cumming. 1944. Reprint, New York: Herder and Herder, 1972.
"The House of Sade." *Yale French Studies*. no. 35 (1965).
Hovey, Kenneth Alan. " '*Montaigny* Saith Prettily': Bacon's French and the Essay." *PMLA* (January 1991).
Hunt, Lynn, ed. *The Invention of Pornography: Obscenity and the Origins of Modernity, 1500–1800*. New York: Zone Books, 1993.
Hunter, C. K. *Paradise Lost*. London: Allen and Unwin, 1980.
Huxley, Thomas Henry. *Life and Letters*. Edited by Leonard Huxley. 2 vols. New York: Appleton, 1900.
Idel, Moshe, "Mysticism." In *Contemporary Jewish Religious Thought*. Edited by Arthur A. Cohen and Paul Mendes-Flohr. New York: Free Press, 1972.
Jacob, François. *The Logic of Life: A History of Heredity*. Translated by Betty E. Spillmann. New York: Pantheon Books, 1973.
Jacobs, Louis. *Jewish Mystical Testimonies*. New York: Schocken Books, 1976.
Jayatilleke, K. N. *Early Buddhist Theory of Knowledge*. London: Allen and Unwin, 1963.
Johnson, Barbara, "The Execution of Billy Budd," in *The Critical Difference: Essays in the Contemporary Rhetoric of Reading*. Baltimore: Johns Hopkins University Press, 1980.
Johnson, Pamela Hansford. *On Iniquity: Some Personal Reflections Arising Out of the Moors Murder Trial*. New York: Scribners, 1967.
Johnson, Paul. *The Birth of the Modern: World Society 1815–1830*. New York: HarperCollins Publishers, 1991.
Jonas, Hans. *The Gnostic Religion: The Message of the Alien God and the Beginnings of Christianity*. Boston: Beacon Press, 1958.
Jonsen, Albert R. and Stephen Toulmin. *The Abuse of Casuistry: A History of Moral Reasoning*. Berkeley and Los Angeles: University of California Press, 1988.
Kant, Immanuel. *The Critique of Judgment*. Translated by James Creed Meredith. London: Oxford University Press, 1952.
———. *Perpetual Peace and Other Essays: On Politics, History, and Morals*. Translated by Ted Humphrey. Indianapolis, Ind.: Hackett, 1983.
Katz, Jack. *Seductions of Crime: Moral and Sensual Attractions in Doing Evil*. New York: Basic Books, 1988.
Keller, Evelyn Fox, "Nature, Nurture, and the Human Genome Project." In *The Code of Codes: Scientific and Social Issues in the Human Genome Project*. Edited by Daniel J. Kevles and Leroy Hood. Cambridge: Harvard University Press, 1992.

———. *Secrets of Life, Secrets of Death: Essays on Language, Gender, and Science.* New York: Routledge, 1992.

Kermode, Frank, "Adam Unparadised." In *The Living Milton: Essays by Various Hands.* Edited by Frank Kermode. London: Routledge and Paul, 1960.

Kerrigan, William. *The Sacred Complex: On the Psychogenesis of Paradise Lost.* Cambridge: Harvard University Press, 1983.

Kevles, Daniel J. *In the Name of Eugenics: Genetics and the Uses of Human Heredity.* Berkeley and Los Angeles: University of California Press, 1985.

———and Leroy Hood, eds. *The Code of Codes: Science and Social Issues in the Human Genome Project.* Cambridge: Harvard University Press, 1992.

King, Martin Luther, Jr. *A Testament of Hope: The Essential Writings of Martin Luther King, Jr.* Edited by James Melvin Washington. New York: Harper and Row, 1986.

Klossowski, Pierre. *Sade mon prochain, précédé du Philosophe scélérat.* Paris: Seuil, 1967.

Koestler, Arthur. *The Ghost in the Machine.* New York: Macmillan, 1968.

Konner, Melvin. *The Tangled Wing: Biological Constraints on the Human Spirit.* New York: Holt, Rinehart, and Winston, 1982.

Koyré, Alexandre. *From the Closed World to the Infinite Universe.* Baltimore: Johns Hopkins University Press, 1957.

Krimsky, Sheldon. *Genetic Alchemy: The Social History of Recombinant DNA Controversy.* Cambridge: MIT Press, 1982.

Lacan, Jacques. "Kant avec Sade." In *Ecrits.* Paris: Seuil, 1966.

La Fayette, Madame de. *The Princess of Clèves.* Translated by Nancy Mitford. New York: New Directions, 1951.

Lecky, W. E. H. *History of European Morals from Augustus to Charlemagne.* New York: Appleton, 1870.

Legman, G. *The Horn Book: Studies in Erotic Folklore and Bibliography.* New Hyde Park, N.Y.: University Books, 1964.

———. *Love and Death: A Study in Censorship.* New York: Hacker Art Books, 1949.

Lévi, E. *Transcendental Magic.* 1896. Reprint, London: Rider, 1962.

Lewis, C. S. *A Preface to "Paradise Lost."* London: Oxford University Press, 1942.

Lewontin, R. C. *Biology as Ideology: The Doctrine of DNA.* New York: HarperPerennial, 1992.

———. "The Dream of the Human Genome." *The New York Review of Books,* May 28, 1992.

———. Foreword to *Organism and the Origins of Self.* Edited by Alfred I. Tauber. Dordecht: Kluwer, 1991.

Lilienthal, Georg. *Der "Lebensborn e.V."* Stuttgart: Gustav Fischer Verlag, 1985.

Loomis, Roger Sherman. *The Grail: From Celtic Myth to Christian Symbol.* New York: Columbia University Press, 1963.

Lovejoy, Arthur O. *The Great Chain of Being: A Study of the History of an Idea.* Cambridge: Harvard University Press, 1936.

———. "Milton and the Paradox of the Fortunate Fall." *ELH* 4 (1937), 161–79.

Lynch, Lawrence W. *The Marquis de Sade.* Boston: Twayne Publishers, 1984.

Macpherson, C. B. *The Political Theory of Possessive Individualism: Hobbes to Locke.* Oxford: Clarendon Press, 1962.

Mallarmé, Stéphane. *Œuvres complètes*. Bibliothèque de la Pléiade. Paris: Gallimard, 1945.

Manuel, Frank, E. *The Broken Staff: Judaism through Christian Eyes*. Cambridge: Harvard University Press, 1992.

Martin, Andrew. *The Knowledge of Ignorance: From Genesis to Jules Verne*. Cambridge: Cambridge University Press, 1985.

May, George. *Les mille et une nuits d'Antoine Galland*. Paris: Presses Universitaires, 1986.

McGinn, Colin. *Problems in Philosophy: The Limits of Inquiry*. Oxford: Blackwell, 1993.

Megged, Matti. *The Animal That Never Was (In Search of the Unicorn)*. New York: Lumen Books, 1992.

Meltzer, Françoise, "The Uncanny Rendered Canny: Freud's Blind Spot in Reading Hoffmann's 'Sandman.' " In *Introducing Psychoanalytic Theory*. Ithaca, N.Y.: Cornell University Press, 1982.

Michaud, Stephen G., and Hugh Aynesworth. *The Only Living Witness: A True Account of Homicidal Insanity*. Updated ed. New York: Signet, 1989.

———. *Ted Bundy: Conversations with a Killer*. New York: Signet, 1989.

Miller, James. *The Passion of Michel Foucault*. New York: Simon and Schuster, 1993.

Miller, Jonathan, "Going Unconscious." *New York Review of Books*, April 20, 1995.

Milner, Richard. *The Encyclopedia of Evolution: Humanity's Search for Its Origins*. New York: Facts on File, 1990.

Monod, Jacques. *Leçon inaugurale*. Paris: Collège de France, 1967.

———. *Le hasard et la necessité*. Paris: Seuil, 1970.

Montaigne, Michel de. *Œuvres complètes*. Biblioteque de la Pléiade. Paris: Gallimard, 1962.

Moulton of Bank, Lord, "Obedience to the Unenforceable." *The Atlantic Monthly* (July, 1924).

Muller, Hermann. *Out of the Night: A Biologist's View of the Future*. New York: Vanguard, 1935.

Nagel, Thomas. *Mortal Questions*. New York: Cambridge University Press, 1979.

Neel, James V. *Physician to the Gene Pool: Genetic Lessons and Other Stories*. New York: Wiley, 1994.

Nicholas of Cusa, Cardinal. *Of Learned Ignorance*. Translated by Germain Heron. London: Routledge and Paul, 1954.

Nietzsche, Friedrich. *The Philosophy of Nietzsche*. New York: Modern Library, 1927.

———. *Unmodern Observations*. Edited by William Arrowsmith. New Haven: Yale University Press, 1990.

Oppenheimer, J. Robert. "Physics in the Modern World," In *Great Essays in Science*. Edited by Martin Gardner. New York: Pocket Books, 1957.

———. "Tradition and Discovery." *ACLS Newsletter* (October 1959).

Ovid. *The Erotic Poems*. Translated by Peter Green. Harmondsworth, England: Penguin, 1982.

Pagels, Elaine. *Adam, Eve, and the Serpent*. New York: Random House, 1988.

Pagels, Heinz, R. *The Cosmic Code: Quantum Physics as the Language of Nature*. New York: Simon and Schuster, 1982.

Paglia, Camille. *Sexual Personae*. New Haven: Yale University Press, 1990.

Panofsky, Dora, and Erwin Panofsky. *Pandora's Box: The Changing Aspects of a Mythological Symbol.* Bollingen Series. New York: Pantheon Books, 1956.

Passmore, John. *Science and Its Critics.* New Brunswick: Rutgers University Press, 1978.

Peirce, Charles S. *Values in a Universe of Chance: Selected Writings.* Edited by Philip P. Wiener. Stanford, Calif.: Stanford University Press, 1958.

Polanyi, Michael. "Life's Irreducible Structure." In *Knowing and Being.* Edited by Marjorie Grene. Chicago: University of Chicago Press, 1969.

Posner, Richard A. *Law and Literature: A Misunderstood Relation.* Cambridge: Harvard University Press, 1988.

Praz, Mario. *The Romantic Agony.* Translated from the Italian by Angus Davidson. 2d ed. London: Oxford University Press, 1951.

Propp, Vladimir. *Morphology of the Folktale.* Bloomington: University of Indiana Press, 1958.

Proust, Marcel. *A la recherche du temps perdu.* Bibliothèque de la Pléiade. 3 vols. Paris: Gallimard, 1954.

Quinones, Ricardo, J. *The Changes of Cain: Violence and the Lost Brother in Cain and Abel Literature.* Princeton: Princeton University Press, 1991.

Raggio, Olga, "The Myth of Prometheus." *Journal of Wartung Institute* 21 (1958): 42–62.

Randall, John Herman. *The Making of the Modern Mind: A Survey of the Intellectual Background of the Present Age.* Boston: Houghton Mifflin, 1926.

Randall, Richard, S. *Freedom and Taboo: Pornography and the Politics of a Self Divided.* Berkeley and Los Angeles: University of California Press, 1989.

Rawls, John. *A Theory of Justice.* Cambridge: Belknap Press of Harvard University Press, 1971.

Report of the Commission on Obscenity and Pornography. New York: Bantam Books, 1970.

Rescher, Nicholas. *Forbidden Knowledge and Other Essays on the Philosophy of Cognition.* Dordrecht: Reidel, 1987.

———. *The Limits of Science.* Berkeley and Los Angeles: University of California Press, 1984.

Rhodes, Richard. *The Making of the Atomic Bomb.* New York: Simon and Schuster, 1986.

Ricks, Christopher. "*Doctor Faustus* and Hell on Earth." In *Essays in Appreciation.* Oxford: Clarendon Press, 1996.

Ricoeur, Paul. *Fallible Man.* Translated from the French by Charles Kelbley. Chicago: Regnery, 1965.

———. " 'Original Sin': A Study in Meaning." In *Conflict of Interpretations: Essays in Hermeneutics.* Evanston, Ill.: Northwestern University Press, 1974.

———. *The Symbolism of Evil.* Translated from the French by Emerson Buchanan. New York: Harper Row, 1967.

Rifkin, Jeremy. *Algeny.* New York: Viking, 1983.

Rosenberg, David, and Harold Bloom. *The Book of J.* New York: Grove and Weidenfeld, 1990.

Rousseau, Jean Jacques. *Reveries of a Solitary Walker.* Translated by Peter France. Harmondsworth, England: Penguin Books, 1979.

Rousset, Jean. *Forme et signification.* Paris: Corti, 1962.

Russell, Bertrand. *Icarus; or, The Future of Science.* London: Kegan Paul, 1924.

Sachs, Curt. *The Commonwealth of Art: Style in the Fine Arts, Music, and the Dance.* New York: W. W. Norton, 1946.

Sade, Marquis de. *Œuvres*. Bibliothèque de la Pléiade. Vol. 1. Paris: Gallimard, 1990.

———. *Justine, Philosophy in the Bedroom, and Other Writings*. Translated by Richard Seaver and Austryn Wainhouse. New York: Grove Press, 1965.

———. *The 120 Days of Sodom and Other Writings*. Translated by Austryn Wainhouse and Richard Seaver. New York: Grove Press, 1966.

———. *Juliette: Œuvres complètes*. Vols. 8–9. Paris: Pauvert, 1987.

Schauer, Frederick. *Free Speech: A Philosophical Enquiry*. Cambridge: Cambridge University Press, 1982.

———. "Causation Theory and the Causes of Sexual Violence." *American Bar Foundation Research Journal* (1987).

Scholem, Gershom. *On the Kabbalah and Its Symbolism*. Translated by Ralph Manheim. New York: Schocken Books, 1965.

———. *Origins of the Kabbalah*. Philadelphia: Jewish Publication Society; Princeton: Princeton University Press, 1987.

Scholes, Robert and Eric S. Rabkin. *Science Fiction: History, Science, Vision*. New York: Oxford University Press, 1977.

Schultz, Howard. *Milton and Forbidden Knowledge*. New York: Modern Language Association, 1955.

Seillière, Ernest. *Le mal romantique: Essai sur l'impérilisme irrationnel*. Paris: Plon, 1908.

Senior, John. *The Way Down and Out: The Occult in Symbolist Literature*. Ithaca, N.Y.: Cornell University Press, 1959.

Shapiro, Robert. *The Human Blueprint: The Race to Unlock the Secrets of Our Genetic Script*. New York: St. Martin's, 1991.

Shattuck, Roger. *The Forbidden Experiment: The Story of the Wild Boy of Aveyron*. New York: Farrar, Straus and Giroux, 1980.

———. *The Innocent Eye: On Modern Literature and the Arts*. New York: Farrar, Straus and Giroux, 1984.

———. *Marcel Proust*. New York: Viking, 1974.

Shearman, John. *Only Connect: Art and the Spectator in the Italian Renaissance*. Princeton: Princeton University Press, 1988.

Sneath, E. Hershey. *Philosophy in Poetry: A Study of Sir John Davies's Poem "Nosce Teipsum."* New York: Scribner, 1903.

Solomon, Robert C., "*L'Etranger* and the Truth." *Philosophy and Literature*. (Fall 1978).

Sontag, Susan. *Styles of Radical Will*. New York: Farrar, Strauss and Giroux, 1969.

Speer, Albert. *Inside the Third Reich: Memoirs*. Translated from the German by Clara Winston. New York: Macmillan, 1970.

Stent, Gunther S. "Limits to the Scientific Discovery of Man." *Science*. (March 1975).

Steiner, George. *Bluebeard's Castle: Some Notes Towards the Redefinition of Culture*. New Haven: Yale University Press, 1971.

Stevenson, Burton Egbert. *Home Book of Proverbs, Maxims, and Familiar Places*. New York: Macmillan, 1948.

Stoller, Robert J. *Consensual S&M Perversions*. Washington, D.C.: American Psychiatric Press, 1975.

———. *Pain and Passion: A Psychoanalyst Explores the World of S&M*. New York: Pleneum, 1991.

Strauss, Leo. *Persecution and the Art of Writing*. New York: Free Press, 1952.

Sullivan, J. W. N. *The Limitations of Science.* New York: Viking, 1933.
Tannahill, Reay. *Sex in History.* New York: Stein and Day, 1980.
Tanner, John S., " 'Say First What Cause': Ricoeur and the Etiology of Evil in *Paradise Lost.*" *PMLA.* (January 1988).
Tauber, Alfred, I., ed. *Organism and the Origins of Self.* Dordrecht: Kluwer, 1991.
———. *The Immune Self: Theory or Metaphor?* Cambridge: Cambridge University Press, 1994.
Thomas, Lewis. "The Limitations of Medicine as a Science." In *The Manipulation of Life.* Edited by Robert Esbjornson. Nobel Conference, no. 19. San Francisco: Harper and Row, 1984.
Tillich, Paul. *Love, Power, and Justice: Ontological Analysis and Ethical Applications.* London: Oxford University Press, 1954.
Tillyard, E. M. *Milton.* New York: Dial Press, 1930.
Toksig, Signe. *Emmanuel Swedenborg, Scientist and Mystic.* New Haven: Yale University Press, 1948.
Trilling, Lionel. *Beyond Culture: Essays on Literature and Learning.* New York: Harcourt Brace Jovanovich, 1965.
Trousson, Raymond. *Le thème de Prométhée dans la littérature européenne.* Paris: Droz, 1964.
Villiers de l'Isle-Adam. *Axel.* 1890.
Vogel, F., and A. G. Motulsky. *Human Genetics: Problems and Approaches.* Berlin: Springer-Verlag, 1979.
Wade, Nicholas. *The Ultimate Experiment: Man-Made Evolution.* New York: Walker, 1977.
Walsh, P. G., ed. *Andreas Capellanus on Love.* London: Duckworth, 1982.
Watson, James D., and John Tooze. *The DNA Story: Documentary History of Gene Cloning.* San Francisco: Freeman, 1981.
Watson, John B. *Behaviorism.* New York: W. W. Norton, 1925.
Watt, Ian. *The Rise of the Novel: Studies in Defoe, Richardson, and Fielding.* Berkeley and Los Angeles: University California Press, 1957.
Weisberg, Richard. *Poethics, and Other Strategies of Law and Literature.* New York: Columbia University Press, 1992.
———. "How Judges Speak: Some Lessons on Adjudication in *Billy Budd, Sailor* with an Application to Justice Rehnquist." *New York University Law Review* (April 1982).
Weiss, Peter. *The Persecution and Assassination of Jean-Paul Marat as Performed by the Inmates of the Asylum of Charenton under the Direction of the Marquis de Sade.* English version by Geoffrey Skelton. Verse adaptation by Adrian Mitchell. New York: Atheneum, 1972.
Wertham, Fredric. *The Seduction of the Innocent.* New York: Rinehart, 1954.
Whitehead, Alfred North. *Adventures of Ideas.* New York: Macmillan, 1933.
Williams, Arnold. *The Common Expositor.* Chapel Hill: University of North Carolina Press, 1948.
Williams, Bernard. *Ethics and the Limits of Philosophy.* Cambridge: Harvard University Press, 1985.
Williams, Emlyn. *Beyond Belief.* London: Hamish Hamilton, 1967.
Willey, Basil. *The Seventeenth-Century Background: Studies in the Thought of the Age in Relation to Poetry and Religion.* London: Chatto and Windus, 1934.
Wilson, A. N. *The Life of John Milton.* Oxford: Oxford University Press, 1983.
Wilson, Edward, O. *Sociobiology, the New Synthesis.* Cambridge: Belknap Press of Harvard University Press, 1975.
Wind, Edgar. *Art and Anarchy.* New York: Vintage, 1969.

Index

Nicholas Rescher

ABOUT THE AUTHOR

■

see p. 50 ftn. for ref. Adam & Eve

Born in New York City in 1923, Roger Shattuck was educated at St. Paul's School and Yale College. During World War II, he served in the Pacific theater as pilot in a combat cargo squadron. After working in the film section of UNESCO in Paris, he held various jobs in journalism and publishing in New York before being named a Junior Fellow of the Society of Fellows at Harvard University.

Roger Shattuck has taught French and comparative literature at Harvard, the University of Texas at Austin, and the University of Virginia. As University Professor at Boston University since 1988, he has made a continuing commitment to the core curriculum.

In 1987, the American Academy and Institute of Arts and Letters gave Roger Shattuck a special award for his writings. In 1990, the Université d'Orléans conferred on him a Doctorat Honoris Causa, and the American Academy of Arts and Sciences elected him a Fellow. He recently helped found the Association of Literary Scholars and Critics, which elected him its president in 1995.